THE BODY DIGITAL

THE BODY DIGITAL

A BRIEF HISTORY OF HUMANS
AND MACHINES FROM
CUCKOO CLOCKS TO CHATGPT

VANESSA CHANG

MELVILLE HOUSE
BROOKLYN • LONDON

The Body Digital:
A Brief History of Humans and Machines from Cuckoo Clocks to ChatGPT

First published in 2025 by Melville House
First Melville House Printing: August 2025
Distributed by Penguin Random House LLC,
1745 Broadway, New York, NY 10019 USA.
www.penguinrandomhouse.com

Melville House Publishing
46 John Street
Brooklyn, NY 11201
and
Melville House UK
Suite 2000
16/18 Woodford Road
London E7 0HA

mhpbooks.com
@melvillehouse

ISBN: 978-1-68589-197-8
ISBN: 978-1-68589-198-5 (eBook)

Library of Congress Control Number: 2025942986

Designed by Beste M. Doğan

Printed in the United States of America
10 9 8 7 6 5 4 3 2 1

A catalog record for this book is available from the Library of Congress

The authorized representative in the EU for product safety and compliance is
Easy Access System Europe, Mustamäe tee 50, 10621 Tallinn, Estonia.
gpsr.requests@easproject.com

To Mum, early adopter, queen of
the PalmPilot, and to Dad, finger on the
repeat button—thank you for showing
me how bodies remember, remix,
and reach for what's next

CONTENTS

INTRODUCTION

I SIT IN A CAFÉ, NURSING A CAPPUCCINO AS I WORK ON this book. My notebook, laptop, and phone are splayed out on the table in front of me, each a channel for different thoughts and interactions. As I scribble in my notebook, my hand moves across the page in the tight loops that trace my cursive. A calendar reminder pops up on my iPhone. With a few deft swipes right, some practiced taps, and a final flick of my thumb, I snooze it for a few hours. Instinctively, my fingers play across my keyboard to return to the open browser on my computer screen. Without thinking, I grab the mouse, slide it, and double-click, and a folder on my computer magically opens. All parts of my daily repertoire, these gestures—writing, swiping, tapping, typing—have retreated to the edge of my awareness.

The influential computer scientist Mark Weiser once wrote that "a good tool is an invisible tool. By invisible,

I mean that the tool does not intrude on your consciousness; you focus on the task, not the tool."[1] By this definition, many of our digital tools seem to have succeeded completely; they liberate our bodies by becoming invisible to users. By closing the gap between our bodies and our virtual selves, touchless technologies such as gesture control, voice recognition, and eye-tracking aspire to channel our pure, natural expression.

Such an interface has long been the holy grail for designers. During the past decade, many interfaces seeking to make this vision reality have entered the market. From the Wii motion console to Leap Motion to the gadgets we all now carry in our pockets, these devices aim to erase the boundary between our bodies and our information. These devices promise a future in which our tools are so intuitive, they vanish. Now, it seems that future has arrived.

These devices—and their supposed invisibility—tell a fundamental fiction: that our bodies remain unchanged by our technologies. Though invisible to our conscious minds, our tools indelibly shape us. I need only glance up from my laptop at my local café to recognize the sloped postures, intent stares, and darting fingers mirroring my own. And the invisible influence on our imaginations and social spaces is even more radical. Technologies are not simply objects but architectures that organize our bodies in space and time.

This book explores how these architectures give form to what I call the digital body: how we feel, move, and become through and alongside digital technologies.

Since Descartes, thinkers have imagined the human body as a kind of machine. In these cultural imaginaries, bodies became the machines of their time. As technologies reconfigured space and time, their logics were projected onto the human body. Clockwork technologies sired mechanical automata that mimicked life, from cuckoo clocks to robotic musicians; in return, humans were understood in clockwork terms. Descartes understood the body as an automaton animated by the soul, a view trumpeted by Enlightenment philosopher Julien Offray de La Mettrie in his declaration that man is a machine. The industrial era saw the emergence of the human motor, like the assembly workers on Ford factory lines who performed punishingly repetitive routines. Today, the metaphors persist—only the machines have changed. Digital technologies gave birth to the quantified self, whose very steps are logged daily by their devices. The human brain is cast as a neural network, and cognition as computation. But while these metaphors are potent, they don't quite capture the complexity of our embodied relationships with machines. While technologies often automate our lives—and in the process, contribute to automating us—we are not simply analogues of our machines. You see, this book is less interested in how machines work than how they work with minds, bodies, cultures, and societies. Technology is not separate from us; it is a medium in which we live, think, and create.

This book tells a story about the human body's long entanglement with technology, a history older than most imagine. Living in the era of smartphones and artificial intelligence (AI), it's easy to think that we're in uncharted waters without a map. Our tools have become so frictionless, so invisible, that we forget their historical origins. Long before algorithms and touchscreens, technologies like writing, musical instruments, and even roads reshaped human life. These transformative tools and systems heralded profound changes in how we interact with one another, how we engage with the world around us, and ultimately, how we live.

Legendary speculative fiction author Ursula Le Guin expansively defined technology as "the active human interface with the material world" and "how a society copes with physical reality."[2] That includes not just smartphones and satellites but food storage, clothing, travel, energy—technology as interwoven into the very fabric of life. We think with our hands, learn with our feet, remember by ear: we exist within an ecology of brain, body, and world. Our interactions with technology are dramas of skin, bone, information, rhythm, and power. Technologies refine, track, translate, and choreograph our behaviors; in so doing, they introduce new ways and new languages of being, feeling, moving, and knowing. As these information architectures become increasingly naturalized and integrated into our lives, they encode a more complex, and often hidden, purpose: to structure our attention, calibrate our desires, and train our bodies.

It is one of the great ironies of contemporary technologies that they automate our bodies to the point of obsolescence. Once self-driving cars, motion sensors, and smart homes extract enough embodied data, they no longer need our bodies to mime our lives. Smart homes sync the thermostat to human rhythms of work and sleep, and turn on the lights even if nobody's home. Robot taxis, bereft of passengers, roam common routes. Our habits become patterns and our movements become predictions. But this book is not simply a damning indictment of ubiquitous technologies. As increasingly personalized technologies permeate our lives, it asks urgent questions: How did we get here? What kinds of bodies do our technologies assume, require, or erase? What's at stake when flesh becomes interface? And how might we redesign our path?

Every interaction with a device is a negotiation of meaning. As our technologies hear, watch, and track us, their designs shape how we express ourselves. How might that design foster rich, human-centered interactions? What would it look like to build systems that sense us—not just surveil us, but understand us? Ultimately, this book aims to imagine a future with technology that is mutually creative and deeply human.

The purpose of this book is thus twofold: first, to reveal how the hidden histories of the body in technology shape our expectations and attitudes toward technology today; and second, to help envision new paths forward that are grounded in care, creativity, and collective agency.

Today's frontier technologies—artificial intelligence, robotics, brain-computer interfaces, and cryptocurrencies—often seem torn from the pages of a science fiction novel. At their most hyperbolic, they promise to address intractable problems and advance human societies. Even as these potent new tools predict disruption, innovation, acceleration, and even liberation, in many ways our future appears uncertain, dire, and worst of all, inhumane. But all technologies are unnerving when they first come along. The advent of paradigm-shifting new technologies such as writing, sound recording, and AI typically arouses a cocktail of anxiety, fear, and optimism. They can provoke such profound changes in how we interface with the world that they alter our embodiment: how we make, how we talk, how we see, how we hear, how we move, how we sense, how we feel, and how we think. They shift the very idea of what it means to be human. By reshaping our bodies, they trigger deep-seated hopes and fears about life and death, being alone and being together, our resilience and survival, the worlds we build and inhabit, and individual and collective memory. That is certainly true of the digital revolution that has defined much of the information age from the late twentieth into the twenty-first century. By tracing how we've historically embraced, resisted, and reimagined new technologies, this book affirms the deep and ever-evolving humanity that animates our engagements with all tools. The digital body is not an abstraction—it is us, becoming, again and again, in the technologies we build and the worlds we inhabit.

This book is a map of the digital body. A map is a guide to territory. It allows an aerial view of how territories fit together, and how the natural landscapes they contain collaborate with built infrastructure—how a road, for example, might rhyme with a river. Maps also reveal the connections between places, suggesting possible routes and relationships. But as cartographers know, there is no perfect map of reality. Maps are representations, contoured by choices about what to include and what to leave out. Maps both depict and direct; they chart terrain while enabling movement through it. This book serves as a map by offering a framework for understanding our richly dimensional and historically situated bodies, which are always emerging from natural, cultural and technological forces. It is also an invitation to imagine and navigate toward a future that is humane, embodied, and alive with possibility.

The digital body is the latest incarnation in an ongoing process of coevolution between bodies and machines. *The Body Digital* traces this entanglement, moving within and beyond the borders of the skin. Each chapter centers on a body part—Hand, Voice, Ear, Eye, Foot, Body, Mind—and explores how technologies have extended, shaped, and transformed it. This history has conditioned how we respond and react to new technologies.

"Hand" opens the story with the gestures of writing, craft, and tapping; by automating performances of the human hand, technologies from player pianos to smartphones translate our most dexterous organ into information and outsource its intelligence to machines. The deep connection between hand and voice—central to human evolution, the emergence of language, and the invention of tools—leads us to "Voice." This chapter explores the haunting and preservation of the self through sound technologies, from the very first voice recordings to vocal deepfakes, and traces the uncanny afterlives of human presence. As machines have transformed vocal communication, they simultaneously transformed hearing. In "Ear," we listen to the evolving acoustics of intimacy and community, from the primeval soundscapes of our first ancestors and our enmeshment in choirs and song, to the freedom and isolation of portable stereo and digital streaming's reconstitution of our social worlds. From sound to sight, the narrative shifts from social networks of listening to human and machine systems of looking. "Eye" shows how cameras and computer vision have extended our gazes and policed our bodies, rendering us both watcher and the watched. "Foot" then considers the intricate choreography of movement and infrastructure, from ancient footpaths to smart cities that sense and guide our steps. While smart cities reconfigure environments as interfaces, immersive technologies do the same to the body itself. "Body" dives into immersive media, where physical form and virtual space blur into augmented experience, highlighting the mixed realities we inhabit every day. The book culminates with "Mind," returning to writing as the first formal technology to offshore memory. This final chapter casts AI as the next stage in the evolution of extended intelligence, redrawing the very boundaries of thought and cognition by projecting it into code.

Though distinct, these chapters intersect and entwine like the systems of a living body. In this ecology, they map the rhythms, ruptures, and reinventions of human embodiment in digital life. Like the technologies they examine, these chapters draw artificial distinctions between nature and culture to illustrate key dynamics. This book is by no means exhaustive; it omits senses like smell and touch, reflecting the uneven ways technology privileges certain perceptions over others. Neither is completeness its goal. Instead, these chapters highlight a recurring pattern: early technologies initiate the separation of actions and perceptions—making, looking, listening, moving, and thinking—from our bodies, outsourcing them to machines. This detachment turns sound, images, and thoughts into objects that can be observed, counted, exchanged, bought, and sold. At the same time, these technologies have turned our bodies into information in various ways—our intelligent hands become data, our faces mapped as geometry, our footsteps counted—all of which can then be parsed, controlled, and regulated by computation. Eventually, these datafied behaviors return to govern us.

We live amid systems capable of replicating, replacing, and transcending almost everything our bodies can do—and in the process, automating our own bodies. And as these systems erode the contexts in which natural human experience once thrived, they also undermine our trust in our own senses.

This history reveals how our quiet submission to early technologies—often in ways that seemed benign—primed our bodies for deeper, more insidious forms of subjugation. But this isn't only a story of control. It's also one of resilience and creativity. Within every constraint, humans have forged connections and nourished new ways of making meaning. It is this latter impulse—our capacity for invention and adaptation—that can help us navigate the technological unknown that lies beyond the horizon of every new digital innovation. After all, machines are never just machines; they are embedded in cultural contexts that shape how we live with them—and how we live with ourselves. Rather than cycling endlessly between fear and hype, let's turn to history and remember when we used machines to tell more imaginative stories, to know ourselves more deeply, to celebrate bodily difference, and to expand what it means to be human.

In that spirit, *The Body Digital* charts territories still in formation. As Ursula Le Guin wrote, "I don't know how to build and power a refrigerator, or program a computer, but I don't know how to make a fishhook or a pair of shoes, either. I could learn. We can all learn. That's the neat thing about technologies. They're what we can learn to do."[3] These histories remind us that in learning how we've built, we discover how we might become.

THE BODY DIGITAL

1

HAND

IN 2018, ARTIFICIAL INTELLIGENCE MADE ITS FIRST LOUD splash in the art world. A trio of French students, calling themselves Obvious, put a smeared, unfinished portrait up for auction at Christie's. Titled *Portrait of Edmond de Belamy*, the murky image pictured a suited gentleman with a plain white collar. Rendered in three-quarter profile against an indistinct background, the image mimicked conventions from Renaissance and Baroque portraiture. At the corner of the canvas, a mathematical equation ostentatiously replaced the traditional artist signature—an overt nod to the algorithm behind its creation. Offered at $10,000, the image went to an anonymous buyer for a jaw-dropping $432,000. Global media trumpeted the achievement as a stunning first. Obvious oiled its promotional engine by claiming it to be the first ever work of art created by AI to go under the hammer. Neither claim turned out to be true: significant human labor went into its making, and a set of AI-generated images had gone to auction at San Francisco's Gray Area in 2016.

Obvious worked with an artificially intelligent system known as a generative adversarial network, or GAN. An algorithm composed of two parts, a GAN reproduces an agonistic relationship between artist and critic. The generator creates new images based on a massive dataset, and the discriminator evaluates that image against human-made images. Obvious trained their GAN on a dataset of fifteen thousand portraits from the fourteenth to the twentieth century. Ensconced in a gilded frame, and gorged on the Western canon, the image convincingly reproduced the aesthetic conventions of portraiture. By signing the image with a segment of the algorithm's code, Obvious had, in a clever sleight of hand, cast an algorithm as a proxy for the absent hand of the artist.

But the artist in question was not a hypothetical person. Obvious had appropriated both the code and its training set from Robbie Barrat, a nineteen-year-old artist and programmer who had shared his Old Masters GAN on Github, the open-source sharing website, in 2014. The auction triggered an outcry from the small but growing community of AI artists, who not only condemned the erasure of Barrat's contribution but also balked at the image itself—an amateur-ish work now held up as the public face of their years of creative experimentation with GANs. In their rush to crown the algorithm as the work's author, Obvious obscured the central roles played by humans in the conception, coding, and curation that yielded the image. If this was a Turing test, then we willfully failed it. But why?

The equation at the traditional site of the artist's signature offers some clues.

Signatures boast a special connection to the human hand. Historically, they have imprinted the heft of the human body upon an object, whether artwork or legal contract. They index the identity of the signer: they affirm that someone was really there. A hand that can sign is a hand that can write, and a writing hand is trained in one of the most significant technologies in human history. Writing embodies an ever-shifting partnership between hand, tool, and mind as they meet world. As a technology, writing highlights how curious and intelligent our hands are.

By displacing a human signature with an algorithm, Obvious culminates centuries of the human hand's slow transmutation by mechanical and computational systems—from craftwork to clockwork to code. This chapter traces that evolution, exploring how the human hand, long a source of embodied intelligence and creative expression, has been mechanized, encoded, and reconfigured. Contemporary technologies, especially those powered by machine learning and predictive computation, extend this transformation into new terrain. From signatures to swipes, they do not merely mimic the gestures of the hand; they anticipate, archive, and even overwrite them. Consider how a smartphone autocompletes your sentences before you've even typed them. The digital body emerges in this dance of movement and machinery, between autonomy and automation. And while AI might seem to be light years beyond writing, both are technologies rooted in the human hand, where mind and body unite in expression.

Writing is most obviously identified by its material legacies: ink scrawled on notepaper, neatly typed rows in a book, or text shimmying across a screen. It can be easy to forget that writing is an embodied activity. Children learn writing alongside hopscotch and jump rope. For my preschooler, these physical activities nourish the same body-mind connection that makes writing possible. When I first learned how to write English letters and Chinese characters, I repeatedly traced meaningless strokes that would coalesce into words, then sentences, then paragraphs, and eventually notes, letters, essays, and stories. Writing disciplines your hands as well as your mind; in return you gain access to new galaxies of expression and to the minds of others, for to write is also to read.

While much of writing's profound impact lies in its massive capacity to store and transmit ideas, its ancestry in handmade marks makes it the twin of drawing. Graffiti artists remind us of this deeply physical connection. As they scale buildings, these renegades turn cities into writing surfaces. They leave tags as tall as they are, still texts that are the residues of human performances. These residues have left an imprint all over human history, evidence of how long our hands have been in conversation with the material world. Prehistoric cave paintings from Indonesia to Argentina, Australia to America, are replete with hand motifs. Some of these handprints simply cohabit with renditions of wild boar, deer, mastodons, and other ancient fauna. In others, these handprints form the bodies of prehistoric menageries. Like graffiti, this rock art captures the hand as an organ reaching out to mark the world—and being remade in return.

I took cartoonist Lynda Barry's Writing the Unthinkable workshop in the summer of 2016. During the weeklong workshop, Barry led her students deep into their memories through their own physical gestures of writing and drawing. As the week went on, the action of writing and drawing seemed to open a portal into my memory. I remembered wood grain under my feet and curlicues in wallpaper, the heat of a summer day searing my skin, and the oily scent of a diner. With incredible clarity, I recalled details, some moving, some mundane. These details became creative resources for the stories, first autobiographical, then semi-fictional, that we told in timed bursts of writing or drawing. Rehearsing these movements (spiralling, scribbling, rapidly sketching) as she urged us to be in our bodies, granted access to what Barry calls the "back of the mind." Before the "front of the mind's" leap to judgment, its preemptive cry of failure and not-good-enough, these movement practices would unleash a visceral surge of images—sensory impressions ripe for the picking.

Barry's method is an expression of haptic creativity that, for many people, digital technologies foreclose. In her telling, something about the experience of computing denies access to that most profound and most prosaic of creative states. She often describes the stalled journey of writing her second novel as a process that, at first, involved a lot of staring at a blank screen. When she decided to ditch the computer and write it with a paintbrush, it poured out of her with a life of its own. Barry's work reveals how deeply our hands mediate the encounters of mind and world. As organs that extend consciousness into our surroundings, hands might be understood as the original interface—or as Barry calls them, "the original digital device"—between human and world.

Paleoanthropologists, neuroscientists, and philosophers have stressed the evolutionary symbiosis of hand and mind.

The hand mediates the most complex interactions of the human brain and the realm of technology. At the same time, our gestures have been shaped by an ongoing dialogue with our tools and our environments. As our earliest principal technology for information storage and retrieval, writing embodies this interplay.

Hands are smart. Hands are curious. Hands learn. Hands know things.

Like many people during the COVID-19 pandemic, I took up making sourdough, among other self-improvement efforts short-lived and otherwise. Making sourdough bread is a long game: a dough goes through several stages of fermenting, stretching and folding, shaping, and rising before it is baked. Knowing when a sourdough has fermented enough or developed enough strength requires a tactile familiarity with the process. How a dough stretches, its texture and pliability, its puffiness, and its spring all mean something to a practiced hand. While I watched several YouTube videos before my first attempt, the reality of nurturing a dough into a crusty loaf was very different from the picture-perfect renditions onscreen. Despite the yawning caverns that pocked that loaf and its less-than-inspiring rise, it tasted pretty good. Since that first go, my weekly rituals have educated my hands in the ways of sourdough. I can gauge readiness with a touch, and my crumb is a revelation.

Surgeons, chefs, and musicians alike gain and exercise knowledge through their fingers. They intuit by touch, sense by weight and pressure, apply force and release, and know when to press, push, pull, cut, or play. This is the "intelligent hand" at work. Imbued with tactile intelligence, hands accumulate experience. Practice embeds this knowledge deeper into their fingertips. These craftspeople, like all of us, practice a repertoire of learned gestures through which they discover and mold their surroundings. Rather than a forceful imposition of ideas onto inert matter, the gestures of craft are themselves conversational encounters with the material world and emblematic of what the sociologist Richard Sennett calls "the evolutionary dialogue between the hand and the brain."[1] Techniques, Sennett writes in his study of craftsmanship, emerge when the hands act as intelligent explorers: probing the world, sending feedback to the brain and acting upon it, gaining expertise through trial and error. Yet, perhaps most surprisingly, such craft practices are also where automation begins to erase the human hand.

Despite the crucial role hands have played in the development of new technologies—and our bodies with them—there have been numerous attempts to automate the human hand out of the equation. A number of ancestors to the modern computer were mechanical objects that reproduced, and eventually effaced, the labor of the human hand. Automata, proto-robots built to act as if working under their own power but actually following a predetermined sequence of operations, have existed for over a millennium. Many of them are dedicated to mimicking the unique human performances of the hand, although they haven't reproduced its intelligence. While most surviving automata are from no earlier than the sixteenth century, tales and visual depictions of automata go as far back as ancient Greece, spanning Europe and Asia. In the twelfth century, Muslim polymath and inventor Ismail al-Jazari served as chief engineer at the Artuqid Palace in Mesopotamia, now known as Turkey. Under this remit, he built a plethora of marvels, including a robot that served drinks, water-operated peacock automata, an elephant clock, and even musical automata.

Over the centuries, many automata have similarly replicated the learned skills of the human hand. The mid-fifteenth-century integration of the steel spring into mechanical design sparked the golden age of clockwork automata, spawning many counterfeits of life. The era witnessed the creation of marvelous clockwork wildlife such as an artificial duck that ate and defecated, as well as some truly remarkable simulations of human artistry and expertise. Of these feats, artificial writing proved a particularly intriguing nut to crack: a mechanical twin to a uniquely human craft. Writing machines sought to both emulate and erase the human hand.

The Writer, designed and built in the 1770s by Swiss-born watchmaker Pierre Jaquet-Droz, his son Henri Louis, and Jean-Frédéric Leschot, was one of these machines. On its face, the automaton resembles a doll come to life. A petite, barefoot boy perches at a gleaming mahogany desk, holding a goose quill. A cloud of brown hair frames his cherubic face, in which crystalline blue eyes dart back and forth, following his own hand as he dips his quill in an inkwell and draws it across a piece of paper. His artificial hand effortlessly controls the pressure of the quill against the sheet, modulating its stroke to achieve elegant, fluid cursive. The boy's red jacket opens at the back to reveal a complex clockwork mechanism comprising six thousand gleaming parts. Even more remarkable, unlike most automata of its day, The Writer is programmable: blocks in his back can be shifted to spell any word and sentence of up to forty characters over four lines. Because of this ability, The Writer is sometimes characterized as an early computer. And indeed, the boy's measured scrawl is the direct ancestor of the mathematical signature at the bottom of Obvious's AI painting.

The Writer hails from a family of three eighteenth-century doll automata, still operational and on view at the Musée d'Art et d'Histoire Ville de Neuchâtel, in Switzerland. Known collectively as the Jaquet-Droz automata, these mechanical marvels—The Writer, The Musician, and The

Draughtsman—reenact three quintessentially human crafts: writing, music, and drawing. Their lifelike gestures reveal not only a fascination with mimicking human skill, but also early anxieties about authorship and embodiment, about what is gained and what is lost when machines perform acts of creative expression. Technological marvels that could recreate human performances, these robots forecast a future in which not only the hand but also its thinking faculties are outsourced to machines.

A few decades later, Swiss watchmaker Henri Maillardet built another writing automaton, his Draughtsman-Writer. It has not survived the intervening two hundred years unscathed. Ruined in a fire, its clothes in tatters, the Draughtsman-Writer arrived in shambles at the Franklin Institute in Philadelphia as a donation in 1928. Once repaired by a staff machinist, rewound, and set in motion, the automaton's florid penmanship produced embellished poems in cursive and drawings of a Chinese temple, a three-masted ship, and Cupids frolicking. With four drawings and three poems encoded on its brass cams (the precisely shaped disks that control its movements) it is believed to have the largest cam-based memory of any such machine of its kind. In fact, it is this very memory that testified to its own origins: after restoration, in a final flourish, the automaton signed, "Écrit par L'Automate de Maillardet" (Written by the Automaton of Maillardet). Storing writing and drawing performances as information, these automata foreshadow a computational future that tames the intelligent hand. In this future, bodies do not simply contain but become data, first by analog and then by digital means. In this way, these automata mark an early step down the long and sometimes sinister road of informatic existence, where the boundary between human skill and machine execution grows ever more diffuse.

How do bodies become information?

In 1804, a French weaver patented a different kind of automaton that mimics and would eventually replace the intelligent hand. Named for its inventor, Joseph-Marie Jacquard, the Jacquard machine is an oft-cited ancestor in the history of modern computing. Fitted to a handloom, it is a mechanical surrogate for the weaver's hand, a physical addendum to the weaving apparatus that automates the production of elaborately patterned fabric.

For centuries, until its invention and widespread deployment in the textile industry, colorful, richly patterned fabric was woven primarily by hand. Drawlooms allowed artisans to manipulate individual warp threads and incorporate different colored weft threads, enabling the production of complex designs. But the process was labor-intensive, demanding years of experience, embodied knowledge and skill, intense focus, and collaboration. Multiple skilled weavers, assisted by a drawboy, had to perform a complicated sequence of gestures so refined that they became second nature.

Once attached to a loom, the Jacquard machine transformed the practiced knowledge of the weaver's hands into a cold choreography of levers and holes—an early expression of computational logic that cleaved skill from the body that bore it. Using punch cards, the machine encodes visual designs as a sequence of instructions. Each card controls a set of narrow, circular metal rods that in turn lift warp threads in response to the presence, or absence, of holes. A hole is an "on" switch and will lift a rod; no hole means "off" and will leave it where it is. Wherever a hole allows it, a rod reaches through and picks up a thread. Stitched together, these perforated maps can produce exquisitely intricate patterns, effectively storing and executing machine-readable data. In this way, the Jacquard loom pioneered binary code and stands as a direct ancestor of the digital languages that power computers today.

By transforming the competence and creativity of the weaver's hand into programmable code—ultimately supplanting that human expertise—the Jacquard loom became the first numerical control machine. The automated innovation allowed less skilled labor to produce complex designs that once required years of mastery, and many artisans soon found themselves obsolete. In this regard, the weaver joins hands with the writers and designers of today, whose carefully honed craft with word and image is under threat of displacement by generative AI. As one of the first AI-generated images to capture the popular imagination, Obvious's *Portrait of Edmond Belamy* echoes the first tapestries woven by the Jacquard loom.

Charles Babbage came of age as a mathematician and inventor as the Jacquard loom was revolutionizing the textile industry. Often hailed as the "father of the computer," Babbage envisioned calculating machines that could mechanize not just labor, but thought itself. In the early nineteenth century, mathematicians, engineers, bankers, and others had to rely on printed mathematical tables to perform large calculations. Incredibly tedious to produce, these tables were often riddled with compounding human error in calculation, transcription, typesetting, and printing. Frustrated by miscalculations endemic to these tables, Babbage set out to eliminate human fallibility from the process altogether.

Babbage's first attempt, the Difference Engine No. 1, was designed to calculate a series of numerical values and then print the result automatically. A section of it, representing one-seventh of Babbage's design, was built by his engineer Joseph Clement in 1832 and comprised about two thousand precision components—rods, ratchets, pinions, and brass gear wheels. Two years later, Babbage conceived of the far more ambitious Analytical Engine, the first design for a general purpose, fully automatic calculating machine. Much like the Jaquet-Droz automata, Babbage's machines housed human reasoning in clockwork: gleaming cogs, wheels, and levers enacting logical operations. Though never built in his lifetime, his Analytical Engine contained the core principles of programmability and automation that underpin modern computers.

Inspired by the Jacquard loom, Babbage designed the engine to be programmable using punch cards. Mathematician Ada Lovelace, Babbage's collaborator and the first computer programmer, articulated the poetic resonance of this design, writing, "The Analytical Engine weaves algebraic patterns, just as the Jacquard loom weaves flowers and leaves."[2] Lovelace's reflection attests to the deep kinship between modern computation and traditional craft, between symbolic abstraction and embodied intelligence. But Lovelace could not have foreseen the magnitude of the transformation these technologies would catalyze. Like the Jacquard loom before it, the modern computer is both a tool of innovation and a harbinger of displacement, opening new creative horizons while threatening to sever the link between hand and mind.

Punch cards were used for computerized tabulation for nearly a century before the dawn of modern computing.

Seeking an efficient, elegant solution to collecting and analyzing census data, American statistician and inventor Herman
Hollerith devised an electro-mechanical tabulator for the 1890
Census that read information encoded on punch cards. The
original Hollerith punch card was approximately the same
size as the U.S. dollar bill so as to fit into boxes made for the
Treasury Department. His system revolutionized tabulation:
it dramatically reduced processing time while increasing the
quantity and complexity of statistics that could be gathered.

Hollerith's success in automating the census data paved
the way for the rise of corporate data empires. His invention
birthed the Tabulating Machine Company in 1896, which
evolved into International Business Machines Corporation—
yes, that IBM—in 1924. While other companies introduced
their own cards, IBM's dominance of the early data processing industry ensured that its format became an industry
standard. By the 1940s, punch cards were integral to early
computers such as ENIAC (Electronic Numerical Integrator
and Computer), ABC (Atanasoff-Berry Computer), and the
Colossus.

Even as IBM cast itself as a champion of automation, its
midcentury promotional materials subtly acknowledged the
embodied origins of computing. In one advertisement for
IBM's Type 604 Electronic Calculator, a glowing human
hand is overlaid with its mechanical surrogate: vacuum tube

modules arranged like fingers. The tagline reads, "Fingers You Can Count On." More than just a sales pitch, the image dramatized a broader shift: the intelligent hand, once a symbol of craftsmanship, reimagined as a modular, electronic appendage—human labor abstracted into interchangeable, replaceable parts.

As this history shows, punch cards were much more than repositories of mere information. Containing an invisible legacy of the intelligent hand—as well as the history of its displacement by automation technologies—they stored human performances of expertise. The pianola, a self-playing piano that came to prominence in the early twentieth century, made this dynamic acutely visible. Like the Jacquard loom, the player piano translated the living intelligence of the human hand into machine-readable data. Anyone who has seen a player piano in action can attest to its uncanny nature. Once in motion, one can almost imagine an invisible player sitting at the instrument. Piano keys seem to play themselves in cheerful ragtime—a mechanical incarnation of the absent human hands in Obvious's AI-generated *Portrait of Edmond Belamy.* Although now largely a relic of a different time, mechanical player pianos stirred many of the same kinds of questions and debates that AI art does now.

Inventor Edwin Votey built a prototype of his automated player piano system in 1895. It took the form of a

large, wooden cabinet and was powered by suction generated by two foot pedals. It was designed to stand in front
of an existing piano, where mobile fingers at its rear would
orchestrate the piano's keys. Astride the piano, this cabinet
of curiosities contained a small paper roll patterned with
tiny perforations representing notes to be played. As a roll
moved over the tracker bar, a reading device with a row of
evenly spaced holes, a valve would open and trigger a pneu-
matic motor. This would then fire a felt-covered finger on
the external player, causing it to hit the corresponding piano
key. Though these functions soon entered the instrument
itself, with the introduction of the Apollo line of pianos by
the inventor Melville Clark in 1901, these operational principles remained the standard for nearly all roll-operated player
piano systems.

Over the next decade, the machine developed into a
ghostly rendition of human performance. In 1904, German
organ builders Edwin Welte and Karl Bockisch launched
a new kind of player piano, the Welte-Mignon. Otherwise
known as a reproducing piano, these were able to record
and reproduce individual performances by replicating the
dynamics, rubato, and pedaling of living players. In its heyday, music roll manufacturers recorded the performances of
many famous twentieth-century pianists, George Gershwin,
Liberace, Jelly Roll Morton, Myra Hess, and Thomas "Fats"

Waller among them. Specters of performances past, these rolls conjure the apparitions of nearly every major early twentieth-century pianist.

For all its commercial success, the automated instrument also sparked fierce criticism. In 1906, American military march composer John Philip Sousa published "The Menace of Mechanical Music," a screed against the sweeping popularity of player pianos and gramophones. Sounding the alarm about these poor copies, Sousa fretted that these "talking and playing machines . . . reduce the expression of music to a mathematical system of megaphones, wheels, cogs, disks, cylinders, and all manner of revolving things, which are as like real art as the marble statue of Eve is like her beautiful, living, breathing daughters."[3] For Sousa, these creative automata substituted cogs and wheels for the ineffable human soul.

At the same time, Sousa identified their menace as even more insidious, suffocating human creativity at its source. As music became effortless to consume, he argued, it would become meaningless to make. He feared these playing machines would discourage young musicians from practice, stifle amateur culture, and rob fledgling musicians of living inspiration. In his telling, only the impressive virtuosity of live performance could awaken "embryotic Mendelssohns, Beethovens, Mozarts, and Wagners to the acquirement of technical skill, or to the grasp of human possibilities in the art."[4] For Sousa, whose bread and butter was patriotic marches, the looming demise of amateurism would cause "the national throat" to weaken and "the national chest" to shrink, fraying social ties and national identity.

And finally—here's the rub—Sousa was concerned about intellectual property rights. Although recorded music had helped to make Sousa famous, he was paid no royalties when machines played his compositions. At the time, copyright law only protected sheet music, not reproduced sound, allowing manufacturers to profit without paying composers. "Could anything be more blamable," he ranted, "than to take an artist's composition, reproduce it a thousandfold on their machines, and deny him all participation in the large financial returns?"[5]

In these various social, metaphysical, and ethical dimensions, Sousa's indictment anticipates many criticisms currently leveled at generative AI: its empty and soulless mimicry; its theft from artists and its ensuing usurpation of creative occupations; its atrophying of practice and skill; and its threat of genuine community. Just as Sousa dreaded the invention of artificial music, we today fear the advent of AI.

Creaky in their physicality, player pianos seem a far cry from the sleek, networked machines so ubiquitous today. However, the rolls that power their spectral movements form a bridge with the artificially intelligent systems that generate such images as *Portrait of Edmond de Belamy*. After all, with the sleek miniaturization of contemporary computing, it is easy to forget that the early computers of the 1940s and '50s were hulking calculators operated by punch cards. In these pockmarked landscapes—Jacquard cards, piano rolls, data tapes—the automated arts took root.

As William Gaddis decried in *Agapē Agape*, his 1998 swan song on the ruination of art by technology, "There was the beginning of key-sort and punched cards and IBM and NCR and the whole driven world we've inherited from some rinky-dink piano roll."[6] The narrator, a dying man on his hospital bed, desperately races to finish his magnum opus: a history of the player piano. Paralleling his deteriorating body with the decay of culture, he rages against a commercialized world where art is reduced to mere entertainment, and imitation supplants authenticity. Yet, while piano players once alarmed devotees of human craft, they did not destroy live performance. In fact, they are invaluable repositories of cultural history, human performance, and creative possibility—something we would do well to remember at moments of great technological upheaval.

Jacquard punch cards and piano rolls alike distilled the expertise of the human hand into machine-readable information. As Gaddis's invective suggests, technology is often cast as the thief of authenticity, of individuality, and of creativity. This sense of loss has long shadowed conversations about technology—especially, to return to the beginning of this chapter, in the history of writing.

In 2010, more than forty U.S. states adopted the Common Core State Standards, educational benchmarks for English and mathematics that scaffolded an ambitious project in education reform. The requirements did not include cursive, even as they affirmed the importance of keyboarding skills for citizens of the future to come. A national backlash followed; while some of this was overtly ideological resistance to the entire education reform project, other critics bemoaned the devaluation of cursive as a uniquely human skill.

Cursive, a special form of handwriting, signifies identity and individuality as well as the hand's creativity and competence. Knowledge of cursive shores up the connection between hand and mind, fosters individual expression, and cultivates self-discipline. Cursive takes a lot of practice, as many of my fellow millennials can attest. It recruits your whole body along with your hands. Cursive in Chinese calligraphy has its own stylistic abstractions that must be learned because they look nothing like the original character. "Running script," the Chinese name for semicursive script, signals how much our bodies invest in cursive writing, as opposed to "walking script," printing by hand.

I took command-line interface computer classes as a child in the 1980s and joined the Internet in the mid-1990s. At the time, the Internet was a secondary skill; all my notes and exams were handwritten. My friend, one of the top-scoring students in Singapore, had a comically large writer's callus on her finger, a monument to her discipline, anxiety, and ambition. While I was nowhere near as studious, I, too, had a writer's callus. If nothing else, I was fond of writing letters by hand. I still know cursive, the way I know how to swim and to ride a bike. Even so, I have found my own hand atrophied by frequent hours on my keyboard. Of course, I still write sometimes, jotting the occasional Post-it. But my hand gets tired and cramps quickly as my script slides lazily into a scrawl. My writing hand is not as smart or as practiced as it once was.

Historian Drew Gilpin Faust believes that Gen Z's lack of cursive literacy cuts them off from historical memory. If you can't write in cursive, you probably can't read it either. Rich handwritten archives—such as personal letters and family documents—become illegible and inaccessible. History will need translators, perhaps a parent or grandparent; Faust thinks people will have to rely on trained experts to report what history, much of which was recorded by hand, was all about. This is a meaningful loss, reminding us that technologies are bearers of culture too. As player pianos preserve a uniquely embodied facet of human history, so too does cursive. For all the criticism these technologies faced at their emergence, they have come to be rich expressions of human history and individuality. How might contemporary technologies like AI do the same?

To be sure, much is lost with the slow disappearance of cursive—and of handwriting more broadly. Even so, cursive's demise has been greatly exaggerated. In the United States, dozens of states have restored cursive to education requirements. Several Canadian provinces have followed suit. In Western Europe, including Spain, Italy, Portugal, and France, cursive remains a staple of early education. Suspicions about cursive, and everything its loss might mean, are deeply ingrained in the history of writing. Whether doomed or, in some contexts, revived, its fate offers an important lesson: the value of preserving multiple modes of expression, especially those that engage the body as well as the mind. After all, these are not linear, homogeneous histories. Some older forms are worth preserving alongside more advanced ones, particularly those that nourish and awaken our embodied individuality. Writing has always been such a form: a profoundly transformative technology whose relationship to the body has never been static.

Before writing, communication existed in the living present, in the evanescent moments of speech. Words were ephemeral, tethered to the moment and to their speaker. Writing unmoors language from that present and transforms consciousness; by reifying the word, written language allowed us to see our thoughts separated from ourselves. Whereas speech would once dissipate at the very moment of utterance, writing gave thought a new life that persisted beyond the speaker. It allowed someone's thoughts to take shape outside the self, endure across time, and be shared with distant others. Today, as digital technologies remake communication, handwriting is now feted for its proximity to the human body. In an age of screens and automation, handwriting becomes not just a tool, but an emblem: a trace of embodied presence in a disembodied world.

Globally, there were a number of efforts to mechanize writing. In East Asia, these initiatives included mechanical woodblock printing in the Tang Dynasty in China around the seventh century, early movable type in the eleventh century, and Korean brass movable type in the twelfth century. These technologies not only expanded access to texts but also fostered literacy, scholarly exchange, and the preservation of diverse cultural and religious traditions across the region. As it would centuries later in Europe, printing technology had a profound and multifaceted impact on Asian societies by enabling new forms of education, communication, and cultural transmission. When German inventor Johannes

Gutenberg introduced letterpress printing to the West in the fifteenth century, he catalyzed a paradigm shift that media theorist Marshall McLuhan would later call the birth of the "Gutenberg Galaxy," a world remade by the advent of mass communication.

In his 1962 book of the same name, McLuhan argued that print didn't just change how we shared ideas and information, it transformed how we perceived ourselves. "Gutenberg Man," as McLuhan called the subject of print culture, was produced by the material and psychological effects of uniform, repeatable text. Whereas handwriting seemed to flow from the individual hands of the writing body, idiosyncratic and visually identifiable, type forced all writing into uniformity. As philosopher Martin Heidegger insisted in his polemic against type, the typewriter makes everyone look the same.

It might also make everyone move the same, imposing a cultural conformity on the hands that use them. Keyboards promise to discipline unruly fingers—whatever languages their owners speak—into their neat rows. The dominance of the QWERTY keyboard in global computing enforces a particularly Anglocentric choreography for hands across cultures. The QWERTY keyboard made its commercial debut in 1874, when Remington launched the Sholes and Glidden typewriter in the market. As one story goes, the QWERTY layout was designed not with the ergonomics of the human body in mind, but rather around technical constraints: its keys were supposedly arranged to prevent the type bar mechanisms from jamming when commonly paired letters were positioned in proximity. Another apocryphal theory is that gimmicky salesmen could impress potential customers by pecking out T-Y-P-E-W-R-I-T-E-R using just the top row. Whatever choices gave us QWERTY, the format has endured for 150 years, from clunky typewriters to laptops, persisting even in the tiny digital keyboards on our smartphones. And in this case, function has followed form: despite its decidedly unergonomic structure, the keyboard has trained legions of typists. Although other layouts exist, such as QWERTZ and AZERTY, popular in Europe, as well as the more ergonomic COLEMAK, DVORAK and WORKMAN keyboards, QWERTY remains a global monolith.

Yet even monoliths can be moved. Around the world, non-Western computer users have developed work-arounds that not only reimagine the keyboard's relationship with language but also redesign the relationship between hand and machine. In China, engineers confronted the seemingly impossible challenge of interfacing a language of more than seventy thousand characters with a keyboard designed for twenty-six letters. According to historian Thomas Mullaney,[7] both the language and the input had to be radically reconceived to connect Chinese hands and minds with personal computers. Through the development of software that allowed the production of Chinese characters using alphanumeric symbols, computer programmers such as Wang Yongmin enabled millions to join the personal computing revolution. For these users, such software uncoupled input and output; where in English, typing a letter yields a corresponding letter onscreen, in Chinese, typing is akin to an act of orchestration, with complex characters generated from keystroke sequences. These innovations suggest that while interface technologies can ossify how we move and express ourselves, human bodies and minds are resilient and adaptable as they coevolve, like writing, with these tools. When it comes to technology, there are many ways to use our hands that are yet to be imagined.

The history of writing reveals conflicting impulses in the relationship of hands and automation. Even as modern computing has condensed the intelligence, curiosity, and expertise of the hand into information, our hands continue to be important interfaces—and indeed, frontiers—for the future of human craft, creativity, and knowledge. Contemporary technologies, ever more personalized, have become new sites for automating the intelligent hand. And like writing and player pianos, these shiny new tools are paradoxical sites of possibility. On the one hand, they are the means through which bodies are disciplined to conform to systems of knowledge, and on the other hand, they can empower expression, individuality, and cultural memory. Today's digital systems, however, are entwined with the caprices and demands of industry in ways that earlier technologies weren't. They are engineered to capture attention, harvest data, and extract value.

New technologies require us to develop new literacies. By developing such literacies, we train our bodies into habitual choreographies. When you learn to write, you are learning not just symbols but the hand motions that turn lines into letters. When you learn to type, you tether your hands to a keyboard, defining your motions in ways that have neurological and physiological effects. Research shows that writing in print, in cursive, or by typing are each associated with distinct brain patterns and significant learning outcomes.[8] Such studies certainly suggest that taking notes by hand lends to better recall for students, perhaps due to the increased time the slower hand gestures create for mental processing. Additionally, recent research suggests that—echoing Heidegger's critique of sameness—while handwritten letters are associated with the different shapes we make with our fingers, typing's tapping gestures are the same for the letter *e* as they are for the letter *g* or any other key, and as a result, do not support memory by fostering that embodied distinction. How we use our hands profoundly affects how we think.

Digital interfaces impose their own kinesthetic grammar.

When you first acquire a smartphone, the interface is clunky. Each interaction feels contrived, each gesture is an intrusion on your consciousness. But as you rehearse these movements, they become second nature. Like the alphabets your hands write into existence, each of these gestures has assigned meanings. As you achieve fluency with them, these gestures become units of the communication structure you form with your device. When you reach instinctively for your phone, it only takes a few unconscious flicks of your thumb to navigate past the lock screen and into your web browser or messaging app. At the same time, you attain a fluency particular to that brand—when your fingers know an iPhone, it's jarring to use a friend's Galaxy. Autocomplete and its annoying cousin autocorrect, useful scourges on our phones, are accomplices to this shorthand; by completing our quickly tapped words, phrases, and sentences before we even have a chance to write them, these predictive text technologies shape our expressions, conversations, and digital social lives.

This cognitive and physical training enables you to express your individuality. Through practice, your handwriting becomes your own, testifying to your identity. Individuals also have unique patterns when interacting with their personal devices. As you type, your fingers play an idiosyncratic composition of keystroke rhythms on your keyboard. Similarly, the swipes and taps on your touchscreen form a living signature of your movement. The emerging field of gesture biometrics uses these movement signatures in security and other applications in interface design.

Yet, even as it promises to secure our information, gesture biometrics raise urgent questions about privacy, surveillance, and knowledge. Our fingerprints, our DNA, and now our very movements are structured and archived by private corporations with little transparency. As interfaces gather your data, they simultaneously train you in their use. These motions become unique—and trackable—parts of your identity.

As children, yogis, and dancers intuitively know, our minds are embodied. According to cognitive scientists and philosophers in the field of embodied cognition, many elements of human cognition are shaped by concrete aspects of our bodies. These include the sensorimotor system, the perceptual system, and interactions with our environment. In moving, we come to know the world.

What is the shape of that knowledge? While writing emerges from millennia of cultural and technical evolution, many of our interface gestures are being defined by designers and engineers in Big Tech. When our movements are choreographed by corporate interests, the potential effects on our minds run deep.

Certainly, by training our bodies into these movement systems, we learn new ways of communicating with expansive networks of data, knowledge, and people. But they train us to speak in a limiting language that primes our thoughts and shapes how we act. Your movements translate to a ready-made palette of autocompleted words and actions that structure your encounters with the world. By training our gestures, these interfaces integrate our bodies within much larger systems of corporate knowledge and data, automating us to be better consumers.

However exquisite their design, gestural interfaces like that on the iPhone are hardly natural or neutral. These devices choreograph many of our daily movements. According to a 2016 study, the average user touches his or her phone 2,617 times a day. With each stroke, our devices become more a part of us—and us of them. By automating our hands, we automate ourselves.

History, however, reminds us of the hand's abiding creativity. We can allow our devices to automate our hands to the point of obsolescence, like the weavers displaced by the Jacquard loom. Or we can follow the Chinese users who created new modes of keyboard interaction to honor their language and culture. Our bodies must animate our technologies, and not be subordinated to them. By designing interfaces that serve human needs, rather than corporate metrics, we can reclaim the hand's role as a living bridge between mind, body, and world.

2

VOICE

ON MARCH 9, 2008, THE EARLIEST KNOWN RECORDING OF A human voice was played for the first time in 150 years. It released an ethereal and haunting tune crackling with age: a few lines of the French folk song "Au Clair de la Lune." These ten seconds of audio were captured in 1860 by the phonautograph, a device invented by Édouard-Léon Scott de Martinsville that etched sound waves onto soot-covered paper. The recording, never intended to be heard, was played by American scientists who used a virtual stylus to read the contours of the sound waves, reanimating a voice long lost to history.

At the time of the recording it would be another twenty years before Thomas Edison's phonograph, a "talking machine" designed for both recording and playback, stunned and delighted contemporary listeners. Edison himself described his own astonishment when he played the words he first spoke into the phonograph, the children's poem "Mary Had a Little Lamb," later recalling, "I was never so taken aback in my life."[1] It is no surprise that these maiden recordings are of rhyme and song, long used as devices for community memory and its transmission. These grainy recordings capture the human voice on the cusp of profound transformation. Innocent of the future, they croon the singularity of the voice.

Like your fingerprint, your voice is unique. Your vocal signature has a fundamental connection with your body, emerging from the idiosyncratic topography of your lips, mouth, and vocal cords. The subtleties of accent, pitch, and cadence, contoured by your habits, your geography, and your personal and social history, imbue your voice with its identity, which some have called its DNA. Though difficult to describe, these are the features to which the human ear is highly attuned.

Our voices mark our privileged status as a species. After all, we are one of the few animals, including parrots, songbirds, whales, and elephants, who can talk. Our voices bring us into community with one another; they are the foundation of language and the vessel for song. In West African griot traditions, the voice is not just a tool for communication but a living archive—carrying memory, history, and lineage on the breath and across generations. According to ethnomusicologist Gary Tomlinson, singing is intertwined with human history, culture, and evolution. In his telling, "musicking was always technological"[2] and social, insofar as vocal and gestural communication is bound up in the history of toolmaking and use.

Philosophers and poets alike have long seen the human voice as an expression of the soul. In his novel *Hyperion*, Henry Longfellow wrote that "the soul reveals itself in the voice only . . . The soul of man is audible, not visible."[3] Ancient Greek philosopher Aristotle likewise described the voice as "a sound which is the sign of something," produced by "the impact of inspired air upon what is called the windpipe under the agency of the soul in those parts."[4] Emanating from the interior of the body and animated by breath, the voice signifies life itself.

Until the dawn of sound recording, the human voice was tethered to the human body. Speech and song were ephemeral, dissipating in almost the same instant that they sprung into being. At the end of the nineteenth century, sound recording severed voice from body and gave it a new and separate existence, extending what the technology of writing had long before begun. Human voices could now endure beyond death, transcending the limits of the human body. In his 1968 book *The New Soundscape*, composer and acoustic ecologist R. Murray Schafer coined the term "schizophonia" to describe this splitting of a sound from its origins. Intending for it to be a "nervous word," Schaefer characterized modern soundscapes as aberrations in which machine-made sounds subsume natural ones. Sound recording begets a soundscape where sound can be issued from anywhere in the landscape and not simply from a hole in the head.

Like the hand, the voice is a threshold between body and world. Once only borne aloft air, its vibrations now travel wires, waves, and code. If writing extended the hand's reach, sound recording gave the voice a second existence. Translated by machines, abstracted into data, and refigured into new forms, the voice has lived a thousand new lives—pressed into vinyl, remixed by DJs, morphed by Auto-Tune, parsed by speech recognition, and now reanimated by AI-generated vocal clones. These technologies have not only transformed how the human voice sounds, but how it is made, perceived, and preserved. Perhaps that's why sound technologies are rife with fantasies of technological necromancy.

Technologies that reproduce the human voice have always stirred existential concerns. In the 1870s, as Edison was perfecting the technologies that would become the phonograph, he celebrated the instrument's capacity to preserve "the sayings, voices and the last words of the dying member of the family—as of great men."[5] For its earliest listeners, sound recording heralded a new era in which the voice would not die with the body. By preserving the voice, the phonograph promised to sustain life beyond death. For the first time in human history, sound was unfixed from time—a metaphysical shift that was profoundly disorienting. According to an 1877 *Scientific American* article, the machine presented "the illusion of a real presence"[6] that, like contemporary deepfakes, were difficult to discern from human speakers. Just as he denounced player pianos as the "menace of mechanical music," composer John Philip Sousa rued the day these "talking machines" came into being and cast sound recording as a "substitute for human skill, intelligence, and soul."[7] For Sousa, the stakes were incredibly high, with the phonograph threatening to supplant the most intimate of relationships, and in so doing, risking turning a generation of humans into machines, akin to a kind of death: "When a mother can turn on the phonograph with the same ease that she applies to the electric light, will she croon her baby to slumber with her sweet lullabies, or will the infant be put to sleep by machinery? Children are naturally imitative, and if, in their infancy, they hear only phonographs, will they not sing at all, in imitation and finally become simply human phonographs—without soul or expression?"[8]

Culturally, the phonograph would go on to transmit countless specters from beyond the grave. Its eerie capacity to preserve—and to a degree, revive—the dead was famously illustrated in an 1898 painting by Francis Barraud. The image depicts a dog named Nipper listening intently to a gramophone, nose grazing the edge of its horn and ears perked in rapt attention. As the story goes, Barraud inherited Nipper from his late brother and noticed the dog's keen interest in the recorded voice of his dead owner emanating from the gramophone. After rendering the scene on canvas, he sold the painting, copyright, and trademark to the Gramophone Company, which would go on to become music industry juggernaut HMV. The now-iconic imagery became a hallmark of the music industry, adorning records and advertisements worldwide. According to media scholar Jonathan Sterne, these ideas about storing the voices of the dead emerged from an ethos of preservation first popularized through practices of canning and embalming during the Civil War. The phrase "canned music" reflects this uneasy lineage: sound sealed, suspended, and reanimated at will.

But early sound media were never just about mere preservation. Well into the twentieth century, nearly every new communication technology was harnessed in service of cultural fantasies of transmitting disembodied voices from the great beyond. During Spiritualist seances, mediums mimicked the rhythmic clicks of the telegraph to summon spirits. Self-proclaimed "phone-voyants" touted their unique abilities to reach the departed by telephone. In 1920, *The American Magazine* published an interview with Edison during which he discussed his work "building an apparatus to see if it is possible for personalities which have left this Earth to communicate with us."[9] These efforts attest to the longstanding entwinement of science and the supernatural— and of the belief that telecommunication technologies might serve as channels into the afterlife.

Digital technologies opened new spaces for these occult imaginaries. From the invention of the telegraph to the later introduction of television and computers, electronic media have persistently been associated with otherworldly phenomena. Divesting spirit from body is a primal act in much science fiction, after all. The radio play *War of the Worlds* and the TV series *The Twilight Zone*, among many other popular broadcasts, tapped into a deep cultural fascination with how new communication technologies warped time and space. This mythology continues to haunt cultural thinking about cyberspace, virtual reality, and the Internet as ghostly domains that leave our bodies behind.

While communication technologies of all kinds have been haunted by apparitions of the dead, sound media have uniquely borne something of a life essence. Emanating from the interior of the human body, the voice has signified something ineffable, something vital, of a person's identity. Vinyl records, like the sooty paper of the phonautograph and the phonograph's waxed cylinders, contain a trace of the human body. Like a dinosaur footprint fossilized in the earth, these etched sound waves chronicle a moment in time. Residue of the breath, these silent marks, like Scott's in "Au Clair de la Lune," testify that someone was really there. In that instant, they spoke, they sang, they existed. Since the moment of the voice's very first capture, sound technologies have borne the weight of this belief: that within recorded sound lives not just speech, but life—something of the soul.

As these attitudes about the elevated status of the voice shaped sonic media, so too have these technologies remade the voice. Microphones expanded the expressive palette for live vocal performance, allowing singers to explore the quieter reaches of their range. Perched near a singer's lips, the device amplifies even tiny sonic vibrations, the airiest of whispers. Jazz singer Billie Holiday and crooner Bing Crosby, in particular, used the microphone's amplification to usher in a subtler vocal form that cultivated a sense of intimacy with their audiences. Seen as a vessel for the voice's soulful authenticity, microphones are often so embedded in performance that they don't even register as technology. A classic of 1990s pop culture, *MTV Unplugged*, showcased this paradox. The program featured popular musicians performing stripped-down, mostly acoustic sets. Iconic performances included such acts as Nirvana, with Kurt Cobain on microphone, and Lauryn Hill, yes, also on mic. Microphones were standard issue on the series, "unplugged" insofar as they simply transmitted the live voice—a signifier of life itself.

But if live performance centers the voice as a living, ephemeral event, recorded music abets the voice's transformation into a thing—detached from the body, commodified, and endlessly mutable. Sound recording packages vocal singularity into an artistic vision and style, turning it into intellectual property. For this reason, the use—and misuse—of sonic artifacts have been especially charged, sustaining intense criticism and heady celebration.

Hip-hop DJs stirred this controversy by willfully using records as they were never intended in order to create sample-based music. By mixing records on turntables and, later, in studios, DJs and producers placed musical fragments into new contexts—at times, radically changing their original meaning. Detached from their original bodies, records became a medium for newly elastic voices to stretch and squeeze. Slowed down and sped up by dexterous fingers, they stretched vocality to chipmunk highs and rumbling lows. James Mtume, a jazz percussionist who played with Miles Davis, condemned this practice as "artistic necrophilia."[10] This damning phrase highlights the ongoing connection of recorded sounds with their embodied origins—and for some, the transgression of a sacrosanct border between life and death.

These analog manipulations set the stage for even more radical transformations to come. Remaking the very sound of our voices, digital tools such as Auto-Tune erode the boundaries between humans and machines. Auto-Tune is an audio processing application that applies algorithms for pitch correction. The software listens to vocals, and measures and alters them to the exact pitch of the nearest note—correcting for mathematical perfection. Used with a light, discreet touch, Auto-Tune buoys the human voice by disguising off-key inaccuracies, and is imperceptible to the untrained ear.

Cher's 1998 hit "Believe" famously, boldly, and audibly deployed the processor to robotic effect. Cher—then thirty years deep into her career and instantly recognizable to many—distorts her voice into its synthetic double. At some moments, it is a soaring cyborg refrain urging as much as it asks, "Do you believe in life after love"; at others, it becomes a tremulous warble you might hear while on hold with an answering machine. The first hit single to conspicuously use Auto-Tune as an instrument, "Believe" embraced a manifestly artificial aesthetic in an industry that had long lionized the human qualities of the voice. The music video leaned into this technological artifice: Cher, polished by surgery, makeup, and lighting, looks as plastic as the song sounds. In the wake of the song's success, Auto-Tune revolutionized the sound of popular music. Its pop evangelists include T-Pain, Bad Bunny, Daft Punk, among others.

Yet it is not only in the glossy realm of pop culture that these transformations permeate. Virtual voice assistants like Siri and Alexa integrate this robotic vocality into everyday life, revealing how mundane schizophonia has become. Ringing with the unnatural tenor of synthetic speech, voices that never had a body orchestrate many of our interactions—with one another, with our own homes, and with the systems that run our lives. Often coded as women, they evoke the gendered history of secretarial labor. Simulating these familiar human relationships, they are symptomatic of the broader cultural shifts technology has wrought on vocality over the past century and a half.

Filling our lives with synthetic chatter, these machines listen to us as much as we listen to them. In the process of hearing us, they remake our voices as pure data. While early speech recognition research aimed to simulate human reason and language comprehension, the field is now built on the acquisition of data for pattern recognition at a massive scale. Speech recognition software essentially breaks speech down into its component sounds, which algorithms then analyze to predict the most probable word match. That is to say, speech recognition technology treats our voices not as vessels for meaning, but as packages of information to be processed. These systems render voices intelligible through statistics, not understanding. By interpreting human speech and responding in kind, these computational technologies abstract our voices from our bodies. Whereas sound recordings once bore a material trace of the person who spoke, these synthetic voices might feel human but belong less to us than ever. Bit by bit, these computing innovations have introduced a robotic quality into our repertoire, blurring the line between human and machine vocality.

This confusion between humans and machines may feel new, but it predates such digital forms. The early modern era was the heyday of clockwork automata, which are complex mechanical devices designed to automatically follow a sequence of operations. The term "automata" has long been associated with automatic puppets resembling animals or people, such as the Jaquet-Droz automata I discussed in the previous chapter. From the cuckoo clock to a defecating duck, these proto-robots aped life in sometimes astonishing ways.

In 1906, the German psychiatrist Ernst Jentsch theorized that clockwork automata could induce an uncanny feeling due to their unnerving resemblance to living beings. Such beings, he argued, embody a grotesque breakdown between life and its absence. "Among all the psychical uncertainties that can become an original cause of the uncanny feeling," he wrote, "there is one in particular that is able to develop a fairly regular, powerful and very general effect: namely, doubt as to whether an apparently living being is animate and, conversely, doubt as to whether a lifeless object may not in fact be animate."[11] Jentsch cited German fantasy and horror author E.T.A. Hoffman's 1816 tale, in which the protagonist, Nathaniel, falls in love with a lifelike automaton he believes to be human. This confusion, Jentsch claimed, is "one of the most reliable artistic devices for producing uncanny effects" in storytelling.

Father of psychoanalysis, Sigmund Freud, later expanded on this idea through the etymology of the German *unheimlich* or unhomely, defining the uncanny as "the class of frightening which leads back to what is known of old and long familiar."[12] In other words, uncanny feelings are aroused by the familiar cohabiting with the alien, when what we know intimately well becomes estranged.

This logic underpins robotics professor Masahiro Mori's concept of the uncanny valley, which describes the eerie, unsettled sensation a person feels when they encounter an artificial object that closely resembles a human being but falls short. More specifically, it hypothesizes a phenomenon whereby the more human a robot appears, the more empathy an observer will feel—until a perceptual cliff is reached, and empathy gives way to revulsion. The valley in question here refers to that plunge: the moment when artificial likeness becomes too close for comfort, and yet is not close enough. In other words, something is off with automata, robots, and simulations in the uncanny valley; they are neither human nor machine, neither living nor dead. In an age of digital technologies, this effect extends beyond humanoid robots to virtual reality, augmented reality, 3D avatars, and yes, to voices. These disembodied voices, neither fully human nor purely machine, neither living nor dead, haunt our soundscapes with a deep strangeness under the veneer of humanity.

In 2021, a film chronicling Anthony Bourdain's life resurrected the renowned chef and travel documentarian from the dead. In *Roadrunner: A Film About Anthony Bourdain*, he can be heard reading from a letter he wrote to a friend: "My life is sort of shit now," he says in his familiar gravelly voice. "You are successful, and I am successful, and I'm wondering: Are you happy?" Was this a recording miraculously unearthed from his effects after his tragic suicide? Not hardly: the director of the film, Morgan Neville, commissioned the sound bite from a company that uses artificial intelligence to produce vocal deepfakes from sonic archives.

Neville defended this creative choice, but public outrage decried it as a creepy blasphemy of Bourdain's memory, the worst kind of artistic necrophilia.

Despite this outcry, Bourdain's virtual Lazarus is but the latest incarnation of technological reanimation. Reproduction technologies have been used to conjure dead celebrities in the past: famous examples include Tupac Shakur's hologram performing at Coachella and a CGI Fred Astaire tap-dancing through a Dirt Devil vacuum cleaner commercial. Hologram resurrections are almost an industry unto themselves, with high-definition laser and digital technologies conscripted in service of music hologram concerts that merge these crisp reproductions with live orchestration. The aim is to bring these resurrections closer and closer to life. As David Rowell writes in *The Endless Refrain*, for their audiences, these digital performances can foster a visceral sense of connection with an artist's onstage persona, and at their best, make old music new again.

Deepfakes, however, bring such revivals into an uncanny new terrain. As voice clones generated by machine learning systems hew ever closer to their human origins, they become indistinguishable to the human ear. For some, this nascent technology holds great promise—offering realistic vocal models for people with speech impairments, more convincing voice assistants, intimate chatbots, and myriad uses in the entertainment industry. For others, a foreboding future looms where vocal deepfakes erode trust in traditional forms of evidence, and herald even more annoying robocalls and phone scams. However, never ceasing to surprise and amaze, such technologies can certainly be used for good: in late 2024, UK mobile operator Virgin Media O2 brought darling granny Daisy to their team. Daisy is a composite of several AI models that responds to phone scammers as a kind, elderly woman; by rambling on about her cats, her family, and her knitting, Daisy wastes their time, a computational effort that translates into fewer people being scammed. Because of the essential connection of the voice with human identity, these vocal avatars threaten to change the very meaning of being human.

Corporate initiatives in AI voice synthesis have proliferated over the past few years. These systems all basically learn to speak by analyzing and replicating human vocal nuance from massive caches of audio data. Researchers have mined existing audio archives to generate voice clones of celebrities and other public figures. In 2019, a pair of Facebook AI researchers, Mike Lewis and Sean Vasquez, released the results of their speech synthesizer, MelNet. Trained on a 452-hour dataset including more than 2,000 TED talks, the machine learning system generated uncanny vocal clones of Bill Gates, Jane Goodall, and George Takei, among other famous voices.

Similarly, a 2016 project from Google DeepMind synthesized voices by sampling existing human speech. Since then, a number of international start-ups and research groups have continued to develop the technology and its applications in ways that test traditional boundaries of identity. Cambridge-based Modulate builds voice skins that allow you to cloak yourself in someone else's voice. Baidu's Deep Voice can swap a voice's gender or accent. Other projects are more altruistic. Through Project Revoice, a partnership with the ALS Association, Montreal-based AI start-up Lyrebird, named for the Australian bird with the remarkable ability to mimic natural and artificial sounds, aims to restore digital voices to people with the disease who might lose their own.

Like sound recording and Auto-Tune before it, this latest evolution of digital vocality stirs cultural anxieties about the authenticity of the voice, and its capacity to survive death. AI-generated voice clones lie beyond the frontier of the uncanny valley. By detaching human voice from body and turning it into an algorithmic object, sound technologies are more able to indulge in fantasies of immortality. Vocal deepfakes, the spectral afterlife of sound recording and reproduction, animate our voices in the hereafter even as they erase our bodies, our breath, from speech and song.

Musician Holly Herndon, who uses AI as a compositional tool, has invoked artistic necrophilia to describe some contemporary uses of artificial intelligence in vocal deepfakes and generative music. But Herndon does not denounce AI wholesale. Instead, she treats machine learning as a creative medium, pressing it to perform in new and imaginative ways. Her creative approach stands in sharp contrast to projects like OpenAI's Jukebox, a neural net trained on vast datasets spanning music from almost every genre, which pirates original artists by generating songs and lyrics in the style of artists both living and dead. In such systems, machine learning becomes a tool of ventriloquy, pirating the voices and styles of dead performers, and consummating the desire for resurrection that has haunted sound media from the beginning. Used to reanimate the dead, vocal deepfakes raise pressing ethical questions about the agency of the deceased.

For the living, vocal clones can be a means of wresting life from loss. After losing his voice to throat cancer, Val Kilmer partnered with the company Sonantic to create a synthetic voice model trained on his audio archive. Coupled with text-to-speech technology, the resulting voice clone gave him the capacity to speak with his original voice—or something very much like it. But vocal likeness alone shouldn't necessarily verify someone's humanity. Synthetic vocality has been a feature of Augmentative and Alternative Communication (AAC) for decades. The physicist Stephen Hawking, one of the technology's most prominent users, was readily identifiable by his synthesized voice, even making several cameos on *The Simpsons.*

Nonetheless, even as these systems give voice, they also stifle its full expression. In her essay "I Still Have a Voice," disability activist Alice Wong describes the limits of her text-to-speech app for supporting her communication: "The voice options are robotic, clinical, and white. It mispronounces slang and Chinglish, a mix of Mandarin and English which is part of my culture. It also fails to capture my personality, cadence, and emotions."[13] Wong's critique of her AAC tools points to what voice clones might become: a tool that affirms rather than flattens identity, capable of conveying not just intelligibility, but cadence, feelings, culture, and self. This would require the enthusiastic participation of the original speakers in the data collection and encoding, and the integration of cultural context into the process—personal, social, and political meaning—which many current systems sorely lack.

Much of the public outcry over Anthony Bourdain's voice clone stemmed from a shared sense of violation: this crass reincarnation could not be what the man himself would have wanted. In contrast, Kilmer's collaboration with Sonantic to create a voice model that resonated with his own then very-much-alive body suggests a different paradigm for the future of human vocality. Based on the sum recordings of a life, a voice model offers not just a digital artifact, but a form of agency: an opportunity to define that afterlife in advance. The question becomes whether the ghost in the deepfake machine is genuinely our own or just a database zombie. While the legal terrain remains murky, we can look proactively beyond our own deaths and collaborate with AI from beyond the veil.

Here, musicians serve as compelling test cases for reimagining the future of vocality. AI is us; it is human labor concealed. Choices about how to curate and label the data used to train deep learning algorithms are just as central to a legacy as the archive itself. Projects like Jukebox demonstrate how musical corpora—spanning genres, cultures, and decades—feed data-hungry platforms not just sound, but style, phrasing, and feel. Since that first musical waveform was etched into soot, millions upon millions of songs have been recorded. These massive databases offer machine learning algorithms blueprints for reproducing elements both tangible and ephemeral, such as style, tone, and mood. These creative parameters suggest one approach to installing guardrails on vocal data. Artists might choose to reflect on the affordances and vulnerabilities of their recorded selves, not simply to preserve, but to sculpt a version of their voice for the future. This "vocal will" could encompass the entire landscape of their personal archive, mere fragments, or nothing at all. By defining their estate plans to reckon with the more ephemeral qualities of our vocal identities, their own artistic styles, and the possibility of new creative contexts generated by artificial intelligence, artists could fashion a future from beyond the grave. Such a plan would set the terms for posthumous creativity by delimiting how, why, where, and by whom a sonic archive is used as training data.

One approach would be to reject the extractive logic of big data. According to researcher Kate Crawford, AI systems and the industry that creates them are motivated by a ruthless logic that "everything is data and is there for the taking."[14] By this logic, discrete objects—whether mugshot or sound bite—are stripped of personal, social, and political meaning and harvested as raw data in service of profit. Easily lifting these artifacts from databases to serve as training data, AI systems appropriate and erase the human labor that went into their making. So determined, an AI resurrection can corrode a person's living memory with the demands of capital and of infrastructure. In imagining their creative afterlife, someone might approach their sonic archive selectively. Rather than allowing all their data to be fodder for creating vocal models, they might allow only specific parts of their oeuvre that represent eras, themes, or concepts in their biography. A vocal model might then reflect a tightly choreographed slice of a life.

A person might also restrict or envision the contexts in which their digital resurrection appears, whether in terms of genre, purpose, or ethical alignment. They could define the conditions of deployment, forbidding commercial uses, or the voicing of views contrary to their beliefs. They might even preemptively rebuke attempts, like Bourdain's digital doppelganger, to ventriloquize their own words. While legal frameworks in intellectual property and copyright law have worked to keep pace with new technologies, not all jurisdictions recognize postmortem rights of publicity. Still, by planning ahead, people and their heirs can mitigate the transformation of their voices into pure commodities, and infuse their hereafters with creative intention.

For most people who do not have famous voices, this kind of preemptive wrangling might seem irrelevant. But the stakes are real. Kurt Vonnegut once likened artists to canaries in the coal mine—early detectors of the social and technological shocks that others have yet to notice. In an era characterized by the capture and manipulation of data, artists are also especially vulnerable to the public availability of their archives. Song has been central in the evolution of the human voice from life force to data stream. The very first recordings were musical, and music has scored several turning points in this trajectory. We are not all musicians, but many of us have voices, and increasingly we leave astonishing caches of data in the wake of our lives: voicemails, podcasts, video calls. In the century and a half since the plaintive lyrics of "Au Clair de la Lune" were captured on the phonautograph, we have come to generate countless such artifacts of our embodied existence as a matter of course. In the broadest sense, how we allow that data to be used is how we choose to live, even in death. AI speech synthesis may threaten the unique status of the human voice. But it could also help us to find new ways of expressing our humanity.

Just as John Philip Sousa received the arrival of recorded music with alarm, today we hear AI's voice clones with awe and unease. But Sousa's trepidation was misplaced, if not without warrant. Sound technologies have created as many possibilities as they have foreclosed. It's not simply what a technology is that determines our future but how we use it to speak, to remember, and to create.

Technologies, like the humans who create them, will continue to spawn as-yet-unimagined futures. However thoroughly we prepare for that future, these parameters will be tested in the ever-expanding capacity of technologies, AI or otherwise, to generate new social, creative, and commercial terrains. Media history is a library of ghost stories, where echoes of the past haunt the present and script the future.

3

EAR

ACROSS NEW YORK CITY, SINGERS GATHER IN BASEMENTS, museums, gymnasiums, boats, and art spaces. A chorus forms, as strangers assume assigned parts sent by email, along with scores for that day's song. They meet, rehearse for a few hours, and then explode in resplendent harmony. Then they post it to Instagram. The Gaia Music Collective gathers one-day choirs composed of effusive strangers who open their voices to one another in one of the most atavistic forms of musical community. Joined in voice, they recall a time before sound technologies remade our ears, and with them, our heard world.

If the voice is how we reach outward, the ear is how we are reached. Where the last chapter followed the voice as it was captured and remade, this one turns to listening—to how it, too, has been reconfigured by technology, and with it, how we attend, connect, and feel close to one another. Our ears, once tuned by acoustic communities, are now calibrated by machines. From choirs to cochlear implants, music boxes to algorithmic playlists, listening has become a mediated act—private, curated, and data-driven. This chapter traces the transformation of listening from an embodied, social experience into a computational process, and asks what that shift means for how we know and love one another.

For millennia, music was a primal, immediate, and often communal experience. Around fires, in homes, theaters and places of worship, at war and peace, punctuating birth and death, rituals across the globe often included music. Always performed live, song and instrumentation took place in ephemeral moments shared between performer and audience. Before the emergence of musical notation, songs were also a form of cultural transmission. Borne aloft on sound waves, music is living memory embodied in the breath.

To hear is to be touched.

Sound is immersive. It surrounds and fills our bodies. Our ears are but one central part of the experience of hearing. The low end—frequencies that plummet below the threshold of human hearing—still vibrate and infuse our bodies. On dance floors, sub-bass vibrations ripple through skin, muscle, and bone; people hug speakers not just to hear the music, but to become part of it. Listening to someone else sing live is profoundly intimate; it is to be alive in the presence of each other's bodies. A voice—that singular expression of an individual—emanates from a body's interior and enters that of another.

As choral assemblies like the Gaia Music Collective demonstrate, our ears are social organs. They allow us to communicate with one another and to explore and interpret the world around us. While it has historically been central to human survival, our sense of hearing empowers us to participate in two fundamentally human expressions of the voice—speech and song.

Much of the action happens in our inner ear, where sonic vibrations are converted into electrical signals that travel along nerves to the brain. The cochlea, a hollow spiral of bone, plays a vital role in this process of auditory transduction. The cochlea is primed to detect subtle nuances in frequency; unfurled, it might be understood in parallel with a piano keyboard, with different nodes along the spiral allowing the detection of low to high frequencies, which we experience as pitch. Human cochleae have better frequency selectivity than most of our mammalian relatives. While animals like dogs and cats can hear a broader range of frequencies with more sensitivity, humans can distinguish fine-grained differences. This ability is essential not only to perceive speech and music, but to navigating the layered sonic environments of daily life.

Several other remarkable features of the human ear contribute to our ability to hear as a socially attuned, interpretive act. Among these are the sensitivity of human ears to sound pressure (with which we can detect sounds as quiet as a whisper and as loud as a jet engine); their capacity for temporal resolution (detecting rapid changes in sound and very brief sounds); and the deep integration of human hearing with cognitive processes that allow us to parse subtle shifts in frequency, tone, range, pitch, speed, or a sound, as well as sift out specific sounds from background noise, the latter contributing to what is known as the "cocktail party effect." Hearing, in this sense, is not just about sound reception; it is about attention, discernment, and social participation.

Animals—whether human, cetacean, avian, or canine—naturally produce sound as an extension of their bodies. Voice and song emanate from a body's interior to the world outside, supporting and contributing to our sonic communities. Even musical instruments—acoustic ones, that is—are extensions of the human body; horns are prosthetic voices, percussion the primeval encounter of hands and surfaces. Essentially, listening in its most primal form is an encounter between bodies. This acoustic communion is shared by humans and nonhumans alike.

Our ears are highly attuned to the symphonies and cacophonies of the world around us. Whether natural or artificial, our soundscapes tell us about our place in the world—and the state of that world. Historically, our ears have been critical to our survival. They sense and interpret significant knowledge about the world around us, distinguishing to what degree that world is natural or otherwise. Just as our voices increasingly interlace with machines, our ears have become embedded within increasingly synthetic soundscapes.

In parallel with technological development, our soundscapes have evolved. Soundscape ecologist Bernie Krause has spent his life recording the sounds of the natural world, amassing over 4,500 hours of field recordings that compose what he calls "the great animal orchestra."[1] In a tropical jungle at dusk, crickets chirp in a seesaw rhythm, poison dart frogs emit nasal twangs, toads rumble, monkeys wail, parrots squawk, and kingbirds twitter as they flitter across the rainforest canopy. In remote and biodiverse habitats that are untouched by development—from tropics to tundra, forest to savannah—all animals claim their own frequency in the sonic spectrum as a matter of survival. Mammals, such as hippos, elephants, and cats, inhabit the lower ends. Birds, with their immense catalog of song, reside in the middle of the spectrum. Insects and bats sound the highest frequencies. In this way, a species can pick out and identify their own from the complex sonic textures, from warnings to mating calls. A biophony, as Krause calls this layered chorus, is a sonic register of a place's biodiversity.

A particular location's biophony is a living document of its health. Krause once recorded the soundscape of an area called Lincoln Meadow in the Sierra Nevada in California, before and after a lumber company performed selective logging. Selective logging, at least according to the company, is more surgical and thus less impactful to a landscape. When he first recorded the region, its soundscape was rich with birdsong of all kinds. After being logged, to the naked eye the meadow still looked healthy, robust, and full of life. His audio recordings, however, told a different story. The region had suffered a profound loss in biodiversity, audible in the absent birdsong that once textured the forest. His recordings also captured another tragedy: a beaver keening in grief for his mate killed during the logging operation, a plaintive cry that would echo the rest of its lonely days. Krause's work was tragically prescient: in 2017, the soundscape ecologist's home, known as Wild Sanctuary, burned down in a Northern California wildfire, and with it, his recording equipment and the original recordings he'd made over his fifty-year career. In a sad irony, the hollowed-out retreat would become a site for listening to how soundscapes change after a fire.

These losses resonate in the human ear, so attuned to the richness of natural soundscapes. Reporting on the powerful sound of places such as the Kielder Forest in Northumberland in the United Kingdom, wildlife sound recordist Chris Watson describes how some places provoke inexplicably bad and unnerving feelings, yet in others only a few hundred meters away, that sense dissipates.[2] Local guides around the globe confirm Watson's account of this phenomenon, suggesting that spaces that sound wrong or evil have been emptied of animal sounds, defined by an unnatural silence. The human ear can detect this ruin, even if our conscious minds perceive nothing out of the ordinary. The biophony once served as a barometer of environmental safety, danger, and vitality—an acoustic map of our surroundings. Like other animals, humans evolved to read these sonic cues. Our ears have historically been enmeshed in sonic community, bathed in human and animal choruses alike.

For Bernie Krause, this makes the present moment all the more urgent. He urges us to listen, to hear what remains of natural soundscapes. Yet now, many of us are profoundly disconnected from these primeval environments. To immerse ourselves in them is to connect with an ancient moment in the history of the human ear. They promote deep listening at a time when our ears are connected to a constant cascade of sound from our phones, our cars, and our earbuds. While BBC documentaries narrated by David Attenborough depict the natural world as vibrant and untouched, it is disappearing one birdsong at a time. At the same time, for many of us, particularly those living in cities, our soundscapes are overwhelmingly artificial. They are punctuated by airplanes on the flight path (I live under one), car alarms, and garbage trucks. I live near a fire station, and sirens are a regular soundtrack in the urban din. In my San Francisco neighborhood, celebrations are often punctuated by illegal fireworks blasting in every direction, whether for the Fourth of July or the Giants winning the World Series.

Biophonies are waning, and with them, the connection of our ears to the world around us. We live in a social world ineluctably permeated by artificial sound and machine noise. Saturated in constant sound, it is little wonder our ears seek solace in personal stereo. These individual soundscapes—little bastions of personal liberty and choice—grant some ownership over the sonic worlds we inhabit. This freedom comes at a cost: a paradoxical induction into new social systems defined by those very technologies. As with our hands and our voices, personal stereo attunes our ears to new forms of information and, in so doing, fresh ways of being together.

Intelligent jukeboxes, handheld portals to millions upon millions of songs, sit in many of our pockets. With such massive musical archives at our fingertips, it can be easy to forget that music was once primarily live, enjoyed in the company of others. With our devices, we can plug in to a private world tailored to our own tastes. Despite the streamlined glossiness of contemporary digital music technologies, music on demand takes root in the earliest days of musical automation. Automated music disrupts the embodied infrastructure of acoustic community. When a musical instrument began to play itself, bereft of human skill or even the need for it, it opened the possibility of private listening. Mechanical instruments such as music boxes and player pianos, and later, phonographs, prototyped listening as an isolated experience: music for one. Wax cylinders and vinyl records brought performances into homes, and Walkmans and MP3 players condensed whole bands into the palm of your hand and transported them right to your eardrums. Each innovation deepened the segmentation and separation of hearing individuals, gradually disconnecting listeners from collective musical life. Automated music incarnated a new listening relationship: ears coupled not to other bodies, but to machines, forecasting the digitally networked listening cultures we inhabit today.

Musical automata existed in the ancient world centuries before music boxes became widespread. While few created before the sixteenth century still exist, surviving texts from ancient Greece, China, and the ancient Islamic world recount their existence, including a mechanical orchestra made for the Han Dynasty emperor in the third century BCE and al-Jazari's floating boat with a water-operated four-piece band—a flautist, a harpist, and two drummers whose rhythms were programmable. These ancient musical robots paved the way for the eighteenth-century clockwork automata that were the immediate technological precursors of the music box.

Music boxes—some of the earliest forms of music on demand—descend from these more public forms of musical automation. Though now largely a curio found in museums, antique shops, and estate sales, music boxes were a popular household instrument from the early 1800s to the early twentieth century, when the player piano and phonograph rendered them obsolete. Today they are thought to be tabletop trinkets, but in their heyday, music boxes could be as large as a record player or a piece of furniture. Some mimicked the natural world: birdcage music boxes, for instance, housed mechanical birds that would sing in exchange for a coin—a haunting omen of both the commodification of sound and of birdsong's demise in the centuries to follow.

Invented by Swiss watchmakers in the late eighteenth century, music boxes share technological ancestry with automata, cuckoo clocks, and other mechanical facsimiles of life. Early music boxes typically had a comb of steel teeth that were plucked by pins on a brass cylinder, which could be changed for different musical selections. Around 1890, the brass cylinder was replaced by a metal disc. Some of the cylinders were astounding works of craft, with upward of ten thousand pins meticulously hand-placed by nineteenth-century craftspeople so multiple songs could be played from the same cylinder. Essentially, these were tiny, self-playing instruments encoded with extraordinary amounts of quantitative information, much like the player piano would come to be. But unlike the pianola and the phonograph, these were automated performances rather than reproductions of human ones—music without a musician.

The music box marked a turning point in the modern objectification of sound. While not the first automatic instrument, it was the first to be mass-produced and widely marketed. In Europe and the United States, music boxes were coveted consumer items, especially among wealthy consumers. They were also exported internationally, and because they separated listeners from live musical contexts, they were important vessels for cultural transmission. Swiss entrepreneurs even built music boxes that played Chinese folk songs for export to Chinese aristocrats. While listening cultures around music boxes varied, they were consistently sold as marvels that could replace human expertise.

Advertising played a crucial role in shaping this narrative. Companies like Regina, an American mechanical music instrument manufacturer that had, at its peak, over 80 percent of market share, marketed the device as revolutionary. Print ads emphasized privacy, convenience, and sound fidelity—and the erasure of human performers. Anticipating livestreaming's limitless access to musical archives, they promised to play "all your favorite music both classic and popular" as well as "all the best music of all composers of all times."[3]

Some of these ads depict the music boxes at the center of parties—entertainment in lieu of a band. Others showed solitary listeners communing with the machine. One ad boasted that "even the baby finds entertainment in the Regina music box."[4] These scenes chronicle a changing listening culture: music boxes began to divorce ears from other speaking and singing bodies, restructuring listening from a communal act into an insular exchange between individual and machine. In so doing, music boxes began to create the channels for a new kind of hearing that would lead to our digital ears.

Since then, numerous mass-produced sound technologies have nourished and evolved the intimacy between ears and listening machines. Phonographs, gramophones, transistor radios, and later, magnetic tape, made it possible for people to listen to music in the absence of a performer. Several sound recording and storage technologies emerged in the wake of the phonograph's invention. Whereas the music box, as an automated instrument, generated sound on its own, these later technologies recorded and reproduced human performances. Each has spawned new auditory cultures, and with them, consonant reimaginations of the ear. As they transformed the voice from its pure alignment with the human soul to a more machinic object, they transformed listening cultures—and the ear itself.

By turning sound, an ephemeral emanation of the body, into a durable artifact, recording and transmission technologies enabled the construction of sonic archives. According to sound and media scholar Jonathan Sterne, recording is much less about preserving a sonic event as it happens than it is about creating and organizing sonic events "for the possibility of preservation and repetition."[5] In other words, recording is generative: it creates the conditions for new social experiences built around its redeployment in new contexts. That is to say, recording looks toward the future as much as it looks back to the past.

Sonic archives, of increasingly longer length as the technology matured, became grounds for social experimentation and formation, and ultimately, fodder for the algorithmic processes that orchestrate many of our listening experiences today. Vinyl allowed such archives of recorded sound to be brought into private and public places alike, doing away with the need for live performers. But it was the cassette tape that gave consumers the capacity to play with the very material of those archives—sound itself. The cassette tape laid two foundations for the new auditory worlds our ears inhabit now: the mixtape, a homemade compilation of music recorded on a cassette that prefigured the playlist, and the personal stereo, which made private soundscapes possible.

Before the compact cassette tape, sound recording was less a personal act than an industrial one—a process bound to the lab and the factory. Phonographs first used wax cylinders, then vinyl, and gramophones used pressed discs. Early sound recording tools were fragile and limited in length, which stymied curated collections of any length; Edison's wax cylinder phonograph could capture two to four minutes of audio, whereas Emile Berliner's disc recorder, the gramophone, could record up to ten minutes on each side. This was still a boon for ethnomusicologists, who used them as a field recording medium. Vinyl records had a longer length, but were mass-produced, pressed, and distributed by manufacturers like RCA Victor and Columbia. Later innovations included aluminum- and lacquer-coated discs, which improved fidelity, but none of these formats enabled everyday listeners to make curated collections. Compiling and remixing music was not yet within reach of the consumer.

Cassette tapes are derived from larger magnetic tape technology, which did in fact allow for recording too. Before the compact cassette tape, reel-to-reel audio technology was cumbersome, finicky, and required specialized training—primarily the domain of radio professionals and dedicated ethnomusicologists, who lugged this equipment out into the field. For most people, music was something you listened to on vinyl or the radio, not something you recorded. The compression of reel-to-reel magnetic tape recorders into the compact cassette was transformative, making portability and reproducibility key features of musical consumption. The cassette tape was developed and introduced by the Dutch company Philips in the early 1960s, and by the end of the decade the cassette business had exploded. The ensuing cassette cultures had all kinds of social ramifications. Because of their compact size and reproducibility, cassettes empowered individuals and subcultures alike. Punk and underground rock music was smuggled behind the Iron Curtain on magnetic tape, fostering underground networks and amplifying resistance.

They also became a medium for a new kind of social interaction, giving birth to a new cultural form: the mixtape. Like vocal distortion and sampling, the mixtape has roots in hip-hop culture. The same impulses that christened the voice as an object to be remixed and remade brought that same agency to those recording. Mixtapes were originally dubbed cassettes of live hip-hop performances. In the 1970s and '80s, DJs and producers distributed their mixes on tape, artfully matching the beats so one song would flow effortlessly into the next. Fans would copy and distribute these tapes among themselves, forging grassroots distribution networks based on affinity and community. Mixtapes gave voice to community values by echoing the songs and sounds that spoke to and for them.

From its roots in 1970s bootlegging, the mixtape would bloom into what essayist Geoffrey O'Brien called "the most widely practiced American art form,"[6] one that became a global phenomenon. In 1995, at the zenith of the mixtape's popularity, Nick Hornby published his novel *High Fidelity*, later adapted into a cult film. Set in a failing record store owned by a music snob, *High Fidelity* celebrates the art and craft of making mixtapes. As Hornby writes, "To me, making a tape is like writing a letter—there's a lot of erasing and rethinking and starting again . . . You've got to kick off with a corker, to hold the attention, and then you've got to up it a notch, or cool it a notch, and you can't have white music and black music together, unless the white music sounds like black music, and you can't have two tracks by the same artist side by side, unless you've done the whole thing in pairs and . . . oh, there are loads of rules."[7] While there may be loads of them, these rules are unwritten, derived from immersion in mixtape culture. If you know, you know.

Song curation is a love language. Making mixtapes

required a lot of thought and reflection, and above all, time. An act of love, it took patience to plan its aesthetic and emotional arc, and to record it. It was a deeply personal and present expression from one listener to another. Not simply the province of hopeful musicians, the mixtape embodied a deeply private, yet highly social, impulse in technology. In a way, it might be understood to be a precursor of the playlist. As a teenager, I made mixtapes for my nearest and dearest—my boyfriend, my best friend, and one time, my dad. As a medium, the mixtape allowed me to express my love and care, either by sharing my own tastes with someone else, or creating an audio experience that reflected theirs. At my wedding, during her speech, my friend played a song from the mixtape I'd made for her when we were sixteen, after brandishing the handmade liner notes I'd crafted from construction paper. While I pulled songs from other tapes and CDs, some mixtape makers perched by the radio, waiting for a song to be played so they could pounce on that record button. Many a song on a mixtape is beheaded, cut off by fingers just a few seconds too slow. You could also record yourself speaking on the tape, as I sometimes did. In other words, making a mixtape was a gift of time as much as it was a gift of music. It was about listening to the machine, to be sure, but it was also about dedicating your ears to that process as a gift. There was no skipping the duration of a song, as there is now with playlists, hastily assembled on Spotify by your-self (with the split-second click of the "add song to playlist" button), a friend, or an intelligent curator cued to the fancies of your ear. In a culture increasingly defined by frictionless consumption, the mixtape hit a sweet spot: it was empower-ing to be the DJ, and compiling it was just labor-intensive enough to make it feel earned. As writer Matias Viegener puts it, "I am no mere consumer of pop culture, it says, but also a producer of it."[8] When you were making a mixtape, you were, in effect, the algorithm, the living condensation of cultural and interpersonal knowledge gained by knowing and loving someone—and creating something that invited them to know you, and love you, back. In this, the mixtape did what no algorithm could ever do.

As a child of the 1980s, my memories of the Walkman are of blissfully drowning out the cacophony of daily life, making a private world where I got to choose the soundtrack. To be fair, that soundtrack was not all that different from the popu-lar ones broadcast on radio, but it was mine. That canary yel-low Sports Walkman, with its built-in FM radio and rubber grip for running, became a vehicle for autonomy. Plugging my headphones in and jacking up the volume—much to my mother's chagrin—cocooned me in a soundscape of my own design, especially once I got into mixtapes. To be able to do that in our family car, squashed as always in the middle seat between my brothers and the loud din of my large family, was life-changing.

The Walkman was shorthand for the line of portable cassette players released by Sony in 1979. While it went by other names—Sound-about, Freestyle (Australia), and Stowaway (UK)—it is the "walk" in its name that gave the technology its transformative power. The Walkman spurred a meteoric rise in the popularity of cassette tapes. In 1983, four years after the Walkman's debut, cassettes outsold vinyl for the first time.

Like the music box, advertised a century earlier, the Walkman promoted the easy insertion of this little machine into users' lives. And like the music box, it also promised personal liberation, intimacy, and individual control of one's own world. Early Walkman ads pictured the device everywhere, in private and public spaces alike. They depicted a paradoxical social world that straddled private and public spaces. Public spaces became a collection of private soundscapes, each person encased in their own little auditory world. In one ad, people from different walks of life coalesce around a stoop, historically a community gathering place for hanging out and shooting the breeze. In this vision, however, all have a Walkman clipped to their waists, wires snaking to the headphones crowning their heads: the briefcase-bearing

businessman, the woman wearing roller skates and leg warm-
ers, the elderly lady perched at her windowsill, the couple in
matching leather jackets and boots. In large, bold print, the
ad crows, "There's a revolution in the streets." The taglines
are all variations on the same theme, emphasizing the bold
contrast between the device's extraordinary lightness and
powerful audio: "It sounds like it weighs a ton."

Global advertising situated the Walkman in every imag-
inable setting: at the swimming pool, on tropical islands, on
jogs, with friends while riding bikes. The world catalyzed by
the Walkman verified one claim in a print ad: "Look around.
It's happening everywhere." TV commercials from Japan to
Australia celebrated the power of the personal stereo—not just
as a gadget, but as a way of being. Paradoxical in its impact,
it would offer profoundly private and personal experiences
of public space, while also bringing together markedly different people. The Walkman fostered new connections between
people, allowing disparate people to inhabit the same space,
even if they weren't listening to the same music. In a way, even
as these ads pictured a world unchanged but for the insertion
of the Walkman into everybody's lives, they also imagine a
wild freedom to strut, dance, and laugh through life. After all,
anywhere a human can go, a Walkman can go too.

Portable stereo technology existed before the Walkman's
public debut. Similar machines were taken on the Apollo

moon missions, with apocryphal stories of Buzz Aldrin playing Sinatra's "Fly Me to the Moon" on his moonwalk. Sony sent a proto-Walkman, the TC-50, with the Apollo Command Module, along with a music mix, requested by Buzz Aldrin, dubbed to the TC-50's compact cassette format. Until the Walkman hit the market, headphones were heavy and clunky ear cans, to borrow some industry slang, that were essentially speaker drivers mounted onto your ears. The lightweight headphones that populate all these Sony Walkman ads were critical to its commercial success. They made the dream of portable, private listening into a stylish reality, turning the Walkman into a cultural touchstone. The arrival of noise-canceling headphones, with Bose's introduction of the feature in 1989, hastened a kind of auditory quarantine, shoring up the private walls between someone and the world around them.

The Walkman initiated the divorce of the ear from the acoustic communities that had defined musical experience for millennia. It sowed the seeds of a new social world and ushered in the age of personal consumer electronics. The Walkman also ruptured the relationship between ear, body, and environment: plugged into their headphones, a person might be somewhere in body but not in mind, leaving streams of other people to navigate around their body in public space. Even at its inception, its creators were hyperaware of this potential. Akio Morita, Sony's legendary boss at the time of the Walkman's launch, was concerned that the device would be too antisocial. In his memoir, Morita recounts how, when he took the prototype Walkman home, "I noticed my experiment was annoying my wife, who felt shut out."[9] He added two elements meant to mitigate this isolation: dual headphone jacks, and a "hotline" switch that lowered the volume and activated a microphone, serving as an intercom—or panic button?—to the world outside. Indeed, some of the ads highlighted the new intimacy the former feature offered, such as couples locked in a romantic gaze at sunset, ears plugged in to the same unit. In one 1979 commercial aired in Japan, a kimono-clad Japanese man and a Caucasian aerobics instructor giggle and gyrate in lockstep as they share a Walkman. Sometimes it is a unifier, with people sharing a Walkman through a split output port and relishing a secret nobody else can hear. In the end, however, isolation won, and later versions of the Walkman dispensed with the hotline button and the second headphone jack.

That personal audio has alienated us from one another is a bit of a truism. Yet, it also set the stage for a new kind of sonic community in which machines play a much larger part. It made machines a central pillar of our auditory experiences, technology intervening between ears and world. It forecast the ubiquity of personalized music by making it possible to play only the music you wish, at any time, in any place. While the Walkman has since faded into the past, succeeded by the portable CD player, digital audiotape, and eventually, streaming, the world it forecast has come to be.

The Walkman and its descendants—the Discman, the minidisc player, and the MP3 player—did more than allow the construction of private social worlds; they enabled on-the-go access to archives. Cassettes could carry up to two and a half hours of recorded sound on the long end, while audio CDs held about eighty minutes of sound. By compressing sound data and sacrificing fidelity for size, the advent of digital formats such as the MP3 increased this capacity. The original iPod, introduced in October 2001, had a then-impressive 5MB memory that its commercials announced could put "1,000 songs in your pocket."[10] Coupled with its ten-hour battery life, and the features that allowed users to create playlists on the device itself, the iPod marked a new frontier for the ear. The iPod didn't just extend what the Walkman started; it redefined the scale of listening.

Being able to carry a massive cache of sound, which could be curated in real time, deepened personalization as a technical priority. Increasingly large databases turn sound into information, compiled into massive archives that can't be parsed at the scale of human perception, only at the scale of big data. With the migration of so much musical information to the cloud, curation and discovery yield to automatic processes, which now orchestrate many of our ears' most mundane encounters. This technology realizes one future forecast by the music box, insofar as it reinforces personal choice over sound, to be played on demand, at any time, in any place you desire. At the same time, it relinquishes that agency to the algorithm.

Smartphone users, perpetually connected to the cloud, have near-immediate access to almost all recorded music. The 24-7 access to this vast universe of sound might seem to offer our ears boundless opportunities for discovery. This access, however, is shaped by the historical movement toward privatization and personalization ushered by mechanical instruments and personal stereos. Livestreaming services like Spotify and TIDAL refine this trajectory into a kind of perfected individualism, their recommendations and playlists tailored to just your tastes. Their algorithms curate listening experiences that seem to touch the very core of your desires and pleasures: more, always, of the same. In this regard, they are symptomatic of what David Rowell identifies in much "new" music as a deep nostalgia for a supposedly more fun and gratifying past, one fed by consumer capitalism's cynical use of the Internet to quickly retrieve and exploit that past.[11] Yet, it is a relationship restaged between you and the algorithm; through processes of dataveillance, to listen is also to be heard. In this data collection and ensuing curation, algorithms continually remake your ear. On a large scale, these platforms and the computational processes that drive them create listeners, labeling and sorting them into taste buckets. For some, this computational curation is the handmaiden to joy and community, for others, the accomplice to isolation.

Glenn McDonald, former "Data Alchemist" at Spotify, sees algorithmic curation as a pathway to bliss. So understood, this is the logical extension of the mixtape; it is the recognition and ongoing formation of musical communities. Having spent a decade shaping how Spotify curates and recommends music to listeners, McDonald describes the foundation of this process as community. One of the major projects that McDonald worked on at Spotify was Every Noise at Once, a genre map of the world's music. Rendered visually, the color-coded scatterplot situates genre by stylistic proximity and feel so that "down is more organic, up is more mechanical and electric; left is denser and more atmospheric, right is spikier and bouncier." "Singaporean electronic," "dutch dub," "uk dance," and "nordic house" reside in the same constellation, while several galaxies away, "anglican liturgy," "german renaissance," and "Korean contemporary classical" closely orbit one another.

For McDonald, the project was not simply motivated by

Spotify's business imperatives but also by a search for seren-dipity and delight. By identifying—first through the ability of intelligent algorithms to parse sonic similarity and then refined by the human ear—the family relationship between musical genres, this project aimed not only to automate musi-cal curation but also to light avenues to more of what some-one likes. He characterizes the work less as an exercise in data and more as an exercise in finding artist communities, listening communities, and fan communities—the musical communities that emerge and constitute from the social gal-axies abetted by sound recording and distribution tools. In a way, the proximity of genre might be understood as the proximity of communities to one another in terms of taste. Before he was unceremoniously laid off at the end of 2023, McDonald and his team had categorized millions of tracks from over a million artists in 6,291 named genres, from an-ime rock to gujarati garba. In the wake of his layoff from Spotify, McDonald lost access to the internal data that pow-ered ongoing updates to the site, leaving it as an artifact of the project's decade-long history.

The categories emerge from the same data that Spotify uses to make personalized recommendations to its listeners. Powered by artificial intelligence, these recommendations are produced from the interaction of the data layer, shared models, and complex algorithms. The data layer refers to the underlying foundation used by Spotify to drive recommendations; it is an enormous pool of data about user behaviors, preferences, listening history, and interactions with the platform. Shared models describe the collaborative filtering approach the recommendation system employs; this involves creating a map of music based on patterns of user behavior that link users, and track and represent shared preferences among users. When you listen to music, the platform listens to you. Through this dataveillance, its complex algorithms learn where on this map your tastes reside, and it proffers music from those same territories. Little wonder, then, that my recent recommendations are a dog's breakfast of children's songs, Motown, and art rock. I'm sure I'm not the only parent whose top songs playlist includes "Baby Shark." In a recent exchange on my work Gchat, while my colleagues gleefully shared their "Spotify Wrapped" end-of-year listening recaps with remarks on their accuracy, I wryly proffered Kermit the Frog as my top artist.

Spotify is certainly not the only streaming platform out there. TIDAL, Pandora, and others deploy similarly complex algorithms to construct these sonic worlds. For many listeners, there are meaningful nuances and differences between these algorithms that reflect the platform's commitment to artists, listeners, and communities. As expansive as they are, and as much as they purport to represent existing musical communities, these recommendations continue to privatize the ear, paradoxically as they inculcate listeners into defined sonic territories.

I will admit, I love Spotify's recommendations. Sometimes, it feels like the app knows me (loves me?) as intimately as the friends who gifted me mixtapes in the 1990s. Discussions on Internet forums tend to weigh the value of TIDAL vs. Spotify algorithms in similar terms, with users describing how, once the algorithm "gets to know you," the recommendations get better. Perhaps part of the lure of these platforms is the promise of being known, even if that is part of the devil's bargain: your data for a facsimile of love.

Author Jenny Odell sees these offerings as part of the edifice and artifice of the attention economy, an echo chamber designed to sell you to yourself. This parallels Kyle Chayka's critique that machine-guided curation on social media has resulted in the flattening of culture, whereby creators of content—that is, music, movies, writing, news (as we used to call them)—shape their work in ways that fit the demands of media feeds that are always trying to anticipate what we will click on next based on the behavioral data algorithms have gleaned from us. In other words, data leads you to more of the same. Or, as Chayka laments in announcing his final departure from Spotify, the app rends songs from important contexts—a musician's album or larger body of work—and turns it into a "disorganized cascade" of content.[12] Passive listening is the sister of passive consumption. Ultimately, this all makes us passive consumers—a far cry from the active mapping that McDonald envisions in his work as a data scientist.

We are not privy to the feeds of others. In livestreams, we listen to intensely private soundscapes—the digital descendant of the Walkman—that are paradoxically fed by constant connection to musical communities. While the algorithmic playlist descends from the mixtape, its curators are no longer friends or lovers but faceless systems, platforms tuned not by intimacy but extraction, orchestrated by the hollow pulse of capital and code.

Our ears are social organs. Digital ears inhabit systems built on—and masquerading as—longstanding, organic communities. Whereas our ears were once attuned to acoustic communities formed of bodies speaking, singing, and listening to other bodies, they are now also enmeshed in computational networks that signify a flattened social world.

Sound technologies contour our social worlds in ways we often don't perceive. Algorithms don't just operate on platforms like Spotify; they are embedded in our listening devices, from earbuds to hearing aids. They filter our sonic environments, deciding what we hear and how we hear it—quietly engineering what counts as signal and what gets dismissed as noise.

In so doing, they quietly mediate our connections to one another and to the world around us.

Modern hearing devices are coded with algorithms that shape how users who are deaf or hard of hearing engage with other people and their environments. Historically, hearing aids have been worn in the ear, amplifying sound for people with hearing loss. The first hearing aids were ear trumpets that, like the ornate horns on gramophones, physically amplified sound waves directly into ears. In 1898, the invention of electric hearing aids marked a shift toward mechanical mediation. By the end of the twentieth century, digital hearing aids had arrived, transforming sound into data to be shaped by signal processing algorithms. These algorithms do more than amplify; they refine. Digital hearing aids convert received sound into a digital signal, applying signal processing algorithms that improve audibility, support comfort by limiting loud sounds, and help to mitigate the interference of background noise.

Cochlear implants go even further. Surgically implanted, they bypass the damaged inner ear and stimulate the auditory nerve directly, using a signal processor to convert sound into electrical signals sent to the brain. Cochlear implants don't transmit sound as many of us know it, but as data. The brain must learn, slowly, to interpret the digital signal as sound. Tuning a cochlear implant is not simply a medical procedure; it's a process of recalibrating perception. Algorithms that do the work of the human brain, however, can be clunky in their execution. In some ways, cochlear implants try to simulate human hearing, but they can misfire. For example, sound processing algorithms in cochlear implants designed to filter environmental sound in noisy environments can help a deaf user zero in on an individual speaker, kind of like the cocktail-party effect, a phenomenon whereby the brain focuses a person's attention on a particular stimulus. But they sometimes dampen all the sound, including speech.

Cochlear implants, like other digital hearing devices, tend to be tuned to particular social, cultural, and economic values. Because they encode frequency relatively poorly, cochlear implants don't support pitch perception well. As a result, cochlear implants tend to work better for Western languages as opposed to tonal languages such as Mandarin, Cantonese, and Vietnamese. As disability and technology scholar Mara Mills has argued, cochlear implants are coded to prioritize speech over music.[13] In his 2006 memoir *Rebuilt: How Becoming Part Computer Made Me More Human*, Michael Chorost illustrates the impact of this sonic hierarchy on his life. When Chorost experiences total hearing loss, he is especially sad about being unable to hear his favorite song, the classical epic by Ravel, *Boléro*. After getting cochlear implants, he is deeply disappointed to hear the once lush and transcendent song is flat and tinny. *Boléro* becomes the touchstone against which he measures his cochlear implants, and he embarks on a "bionic quest" with his audiologist and other engineers to code its software to allow him to hear its lush orchestration in its glorious depth, color, and tone. But there are trade-offs: when one virtual innovation restores his experience of musical awe, he finds it harder to understand ordinary speech with his implants. The balance between listening modes became a negotiation between communication and beauty.

Today, there is virtually no trade-off between speech and music. Because of software advances, most cochlear implant users often have multiple programs they can toggle for specific listening environments, for example, a designated music program; a designated focus program; a designated program for speech (and in different environments, such as noisy, quiet, and dynamic); and a background noise program. For some users, these choices are welcome; for others, they are fussy and unsatisfying. No longer just a hearing aid, these innovations remake the cochlear implant as a media device—a synthesis entirely in keeping with the evolution of digital ears.

This convergence is not confined to medical technology. Recent years have seen the merging of hearing devices
with headphones—in many ways the realization of a shared
mission to assert personal agency over sonic environments.
It's now common to see people wearing AirPods and other
wireless Bluetooth earbuds that connect to personal devices.
For many cochlear implant users, the device acts as a kind
of Bluetooth earpiece, too, with connectivity that streams
television, phone calls, and podcasts directly to their brains;
the sound is never externalized and remains a purely digital signal from phone to implant. In 2024, Apple released
the AirPods Pro 2, Bluetooth earphones that also function as an FDA-approved over-the-counter hearing aid device. Using adaptation algorithms, the listening device can
block white noise and loud noise while allowing contextual
sounds through. Applied to our sonic environments, these
algorithms are functional versions of the Walkman's claim
to personal space. Such over-the-counter earphones utilize
algorithms designed for assisted listening, whether or not its
user has a hearing impairment.

As design researcher Sara Hendren reminds us, all technologies are assistive—earbuds and hearing aids alike—shaping how we listen, relate, and connect. They mediate not just
sound, but intimacy. As the history of personal stereo shows,
listening technologies can isolate or unite us, sometimes at

the same time. Whether insulating ourselves from the world or drawing us deeper into it, these technologies transmit the emotional logic of the mixtape—carefully curated, intensely personal, designed to communicate feeling across time and distance. How we listen has everything to do with how we know and love one another.

4

EYE

THE MOST FAMOUS PHOTO OF OUR PLANET, *EARTHRISE,* was snapped on Christmas Eve 1968 by astronaut Bill Anders of the Apollo 8 mission. With the craggy surface of the moon in its foreground, the photograph captures half of Earth emerging from the inky black of space. In striking contrast with the luminous lunar landscape glowing in the foreground, the forlorn image of Earth had a profound impact on global consciousness. On the following day, the poet Archibald MacLeish reflected in *The New York Times*: "To see the earth as it truly is, small and blue and beautiful in that eternal silence where it floats, is to see ourselves as riders on the earth together, brothers on that bright loveliness in the eternal cold."[1]

Adrift in endless darkness, our blue marble shone as a

unique refuge for humankind. *Earthrise* approximates what has come to be known as the overview effect, a cognitive shift reported by some astronauts as a sense of awe and transcendence, coupled with a feeling of interconnection with Earth and its inhabitants, upon viewing our planet from space. In 2009, journalist Steve Connor remarked on the abiding impact of this photograph: "They went to the Moon, but ended up discovering the Earth . . . It was an image that would eventually launch a thousand environmental movements."[2] This framing—of the image, as well as its legacy—reflects a fundamentally human perspective of our planet.

And indeed, *Earthrise* was edited for human eyes, or as cultural historian Joe Moran says, "anthropocentric ends."[3] Anders, gobsmacked by the vista of our planet cresting its horizon, took the photo as the Apollo 8 spacecraft rounded the dark side of the moon. He hurried to get color film into the Hasselblad camera onboard, then he snapped three shots through the spacecraft's windows before Earth rolled out of view.

Captive to the views allowed by the spacecraft's trajectory and windows, in its original form, this image did not so clearly convey the uniqueness of our blue dot. The original photo—Earth to the side of the moon, not rising above it—lacked the symbolic drama we now associate with it. According to Anders, "Earth looked so tiny from the heavens that there were times during the Apollo 8 mission when I had trouble finding it. If you can imagine yourself in a darkened room with only one clearly visible object, a small blue-green sphere about the size of a Christmas tree ornament, then you can begin to grasp what Earth looks like from space."[4] Once NASA cropped the image to lift Earth above the moon, the photograph told a more urgent story about the singularity of our planet, and our extraordinary luck at calling it home.

Earthrise is not simply a record of what the Apollo 8 astronauts saw through the hatch window. It is a visual composition, shaped by and for human awe. Where the previous chapter traced how sound surrounds, touches, and connects—how it moves through us—this one shifts its focus to the eye, and to what happens when sight escapes the body, becoming distant, directed, and machinic. As *Earthrise* reminds us, even the most iconic images are not neutral: they are framed, oriented, and curated for meaning. And more often than not, the gaze is no longer ours alone.

Most images of Earth don't tell this story because they're not crafted by or for human eyes. Taken from above, at a vast scale, they instead embody the perspective of machines—and foretell how we would come to outsource much looking to such machines. Abetted by vision machines—first cameras, followed by planes, satellites, drones, and computers—this perspective takes on a distinctly nonhuman quality. This machine aesthetic will eventually come to culminate in the digital imagery that swamps our lives.

In 1982, English heavy metal band Judas Priest released the song "Electric Eye" on their album *Screaming for Vengeance*. With the eponymous electric eye, the song alludes to the omniscient camera in *1984*, George Orwell's dystopian novel, that watched—and policed—the community, updating that camera to a powerful satellite that orbits the Earth. Describing the relentless gaze of the satellite as an enforcement of state power, the song's prescient lyrics forecast modern surveillance society. This chapter tracks the evolution of such electric eyes in the vision machines that, divorced from human perspective, turn on us.

Earthrise was the first color photograph taken of Earth by people from space, and its thoughtful composition reflects that sensibility. While *Earthrise* is commonly thought to be the first photo of our planet, the actual photo first taken of the Earth is far less visually and existentially striking. Earlier images of space attest to the deep strangeness of turning this gaze upon ourselves. In 1946, over ten years before the first artificial satellite, Sputnik, was launched, scientists at the White Sands Missile Range in New Mexico attached a camera to a captured German ballistic missile and blasted it into space. At an altitude of about sixty-five miles, the camera snapped a frame every second and a half. Although the missile plummeted to Earth within minutes, the film survived, offering humans an initial glimpse of the planet. The film chronicles the rocket's dizzying liftoff, capturing a blurry rendition of a part of Earth's curvature and some wisps of cloud. The grainy black-and-white photograph is hardly identifiable. Two triangles form the image: one is the flat black of space, and the other spackled gray field is a section of Earth. Whereas *Earthrise* captured our planet in its technicolor glory, the center of our human story, this early image seems bereft of life, an afterthought. This is a vision machine; at a speed only achievable by machines, it detaches from and outstrips all the limitations of the human body.

Over the following decades, satellites came to colonize low Earth orbit; of the more than ten thousand currently operational satellites suspended in orbit,[5] approximately one thousand are dedicated to Earth observation.[6] These electric eyes now swarm the planet, forming an orbital canopy that sees, filters, captures, and catalogs. Satellite images are now a dime a dozen; Google Earth has made it possible for anyone with a computer to peer at our planet from the skies. Yet it's not even a century ago that the first satellite image was taken, delivering a god's-eye view of Earth. While satellite imagery is now commonplace, its distinctly nonhuman gaze finds its ancestors in one of the earliest vision machines: the camera.

Photography is a technology of detachment that makes

pictures like *Earthrise* possible, that allows the eye to soar. As long as cameras have existed, people have attached these prosthetic eyes to vehicles that can ascend to heights we cannot reach and regions we cannot visit. These flying cameras dramatically expand the field of human vision, even as they liberate that vision from the bounds of the human body. While not all vectors of escape have been machines, they have made nonhuman aerial perspectives available to the earthbound. Balloons, pigeons, planes, and satellites have all been vectors for the eye's flight away from the human body.

Not even half a century before a camera took to the skies on a ballistic missile, cameras escaped the human eye and hand astride birds. In 1907, German apothecary Dr. Julius Neubronner submitted a patent application for an invention: the pigeon camera. Just four years after the Wright Brothers' monumental first flight in 1903, and before extended flights were viable, the pigeon camera offered a perspective that heretofore was unavailable to human eyes. It was just as one might imagine—a small camera strapped to a pigeon's chest, rigged with a delayed-action mechanism. Skeptical that a pigeon could bear the weight of a camera, the patent office initially denied his application, relenting when he produced aerial photographs. Neubronner had been using homing pigeons to exchange prescriptions and medications with a sanatorium since 1903, and his flock were readily available test subjects for his aerial ambitions.

The photographs are skewed in their angles, random in their subjects and framing. Even so, the captured images are remarkable, offering bird's-eye views of land, nature, and buildings in the region near Frankfurt where Neubronner lived; one image is gently framed at its edges by the feathery tips of its carrier's wings. These wing feathers foreshadow the edges of the ballistic missile that blasted a camera into space and snapped the first view of Earth. Wings and rocket tails are both unplanned photographic artifacts of the camera's escape from the human body toward some greater expanse, yet to be known. These artifacts are warnings of the impending erasure of human eyes in the machine vision of today. Developments in aviation rapidly outstripped the need for the pigeon photographers, who continued to serve as message carriers, and people could now ascend the skies and join their avian compatriots in the art of aerial photography. More than a century before drone photography, the pigeon camera indulged the desire of the eye to transcend the Earth.

In the century since, missiles, satellites, and now drones have fostered and naturalized machine vision. As artist Jenny Odell writes, "The view from a satellite is not a human one."[7] The widespread availability of these images highlights how humans have relinquished the politics and aesthetics of photography to machines, as they did to pigeons and ballistic rockets. Humans barely register at this scale. In her artwork *Satellite Collections*, Odell organizes a collection of large human-made objects—parking lots, silos, landfills, waste ponds—cut out from Google Satellite View into a series of prints that highlight their geometry. Satellite images render large-scale structures (that, in reality, dwarf humans) as mere shapes on an aerial map. Even in drone photography, landscape is harvested for its patterns and geometry, rather than the sensual grains of dust at human and hand scale.

One of computing's most powerful aspects is how it can process, depict, and parse vast hoards of data. The machine view is the impossible view: from big distances to big data, this perspective is about making sense at a massive scale. Drone photography competitions capture the most striking visions of this impulse. Flocks of flamingoes descending on salt lakes in Kenya become pink flecks strewn across a luminous surface. A cranberry harvest viewed from above becomes fractal crimson swirls. River tributaries meander and join one another like winding veins on the Earth's skin, humans and animals alike at water's edge dissipating from view. Such drone imagery celebrates a gaze unavailable to the naked eye, one defined by the invisibility of human presence at large scales. As technology has permeated photography, it has come to erase human presence from looking—although the human eye used to be central to the camera's gaze.

Contemporary cameras, as we know them, unfix eye from body. Though now ubiquitous—embedded in nearly every phone and capable of high-resolution, high-focus capture—this estrangement of vision from the body was once inconceivable. Photography's chief ancestor, the camera obscura, relies on the proximity of eye and image. Essentially a pinhole device, the camera obscura projects light through a small aperture into a darkened room or box, casting a live, inverted replica of the world outside—a shadow play of reality.

This optical device has been a tool for artists and scientists alike. Renaissance painters used it to trace nature's contours with uncanny precision, its image preserving ratio and color even as it reversed them. For centuries, they have been used by scientists to study such phenomena as solar eclipses without risking blindness. Physicists from ancient Greece to the Islamic Golden Age experimented with the camera obscura to derive core principles of optics. In eleventh-century Cairo, philosopher Hasan Ibn al-Haytham built a purpose-specific chamber he called Al-Bayt Al-Muẓlim, or the "Dark House," thought to be the first camera obscura, which would inform his magnum opus *The Book of Optics*.

By the sixteenth century, the camera obscura had become

a metaphor for human vision. This analogy defines the relationship between the eye and the seen world by immediacy: just as the outside world is projected onto a darkened room, so too, it was believed, is reality projected onto the eye through rays of light—an image cast upon the body. Photography descends from the camera obscura, turning projection into permanence. Whereas the camera obscura was ephemeral, the photograph imprints projected reality onto a surface, making it durable, portable, and endlessly reproducible. In so doing, it initiated the detachment of seeing from the physical act of looking. Vision, once anchored in the immediacy of the body, became something that could be captured, stored, and transmitted.

Yet, even as both vision and photography evolved into increasingly complex systems, no longer limited to the eye or lens, the metaphor of the camera for the eye endures. The persistence of this metaphor illustrates a deeper paradox at the heart of digital embodiment: we trust what we see, even knowing that sight can deceive. When machines inherit the work of the senses, we transfer that trust to them—forgetting, once again, that the eye has always been fallible. And as developments in imaging technologies have evolved, so too has the digital eye. Today's digital eyes—those of smartphone cameras, Photoshop algorithms, and computer vision systems—do not see as the eye sees, nor do they operate by the same principles of immediacy that the camera obscura once did. They reconstruct, enhance, filter, and infer. As we increasingly outsource seeing to machines, the very nature of sight itself is transformed. Yet cameras, and the images they produce, remain important referents for our reality, even as that reality becomes ever more fluid, manipulated, and abstracted. The digital gaze does not simply record the world; it remakes the very relationship between our bodies and the realities they claim to represent, between what is seen and what is believed.

Photography liberated the eye from the limits of the human body, spawning technologies that capture, magnify, speed up, slow down, and even transform reality. Whereas the eye, like the camera obscura, was once tied to the image of nature, photography detaches that image from the human body, and in so doing, from time and place. Camera technologies allow their subjects to endure, like the recorded voice, beyond death. In so doing, cameras allow our eyes to look at the past.

The eighteenth century incubated a series of experiments and discoveries that culminated in the invention of photography in the early nineteenth century. The oldest surviving photograph was created by French inventor Joseph Nicéphore Niépce circa 1826. He placed a camera obscura on the second floor of his home in La Gras and aimed it at the view outside, projecting it onto a prepared photosensitive plate. He then opened the aperture and exposed the plate to the light. During this protracted period of exposure, light rays etched a blurry, grainy image of the view from the Niépce family's country house, with a pigeon house, the barn's sloped roof, the bakery's chimney, and a pear tree. Bathed in light over a lengthy period of exposure from morning through afternoon, these structures are lit from both sides. Tattooed by light, these plates separate image from eye. Calling it a heliograph—written by the sun—Niépce's invention was the result of a direct positive process, meaning that no negatives were made. In this respect, they were like daguerreotypes, the process invented by Louis-Jacques-Mandé Daguerre that would be announced in 1839 and become the dominant form of photography. The image was the direct result of light marking the treated surface—much as the camera obscura did, and much like the image's relationship with the eye. The precursor to photography as we know it, the daguerreotype was also the first format to fix the human body in an image.

Photographs are essentially recordings of light. In the early days of photography, when the process required long durations of exposure, they were also visual logs of time passing.

Niépce's heliograph took somewhere between eight hours

and several days for the image to leave its mark on the treated plate. Moving bodies—people, horses, carts—needed to come to rest to appear on film, or they would otherwise blur the image. The first known photograph of people is a daguerreotype made by Daguerre in 1838. "The Boulevard du Temple" captures a busy Parisian street. Yet, exposed for five to ten minutes, the image is absent of any trace of the moving traffic that pulsed through the boulevard. In a corner, two blurry figures are barely perceptible: a person stands with one foot solidly on the ground, his other food perched on a box as his shoe is polished.

Photography was born in the crucible of the long exposure, at first an instrument of the still, the static, the dead. The earliest portraits are renderings of people remaining still under the camera's gaze for up to an hour. In the Victorian era, long exposures became an unnerving expression of the period's fascination with death—and with recording media, like the phonograph, as a means of transcending it. Postmortem photography was a genre of portraiture that captured the recently deceased for posterity, sometimes posed with living family members. These memento mori—remember you must die—commemorated the deceased for their living family members and was a popular practice in some cultures well into the twentieth century. To this day, death photographers still work in the cremation ghats along the

Ganges River in Varanasi, India's holiest city. Staged family photos contained dead infants who appear to be sleeping, perfectly preserved victims of consumption, or others lost to illness, sudden death, and the void. Rigor mortis was photogenic; ironically, long exposures meant that the dead, already frozen still, were sometimes rendered more sharply than the living. The image of death is fixed and static, while life is blurry, hard to capture. Early photographic technology was, perversely, better at death than life—an eerie prelude to the ossifying effects of digital facial recognition today. As a result, poses were often stiff and lifeless, mimicries of death: lips were usually set in straight lines, bodies static to hold the smudges at bay. Some studios arranged sitters at a posing table, where they could perch on their elbows, or in awkward special devices with headrests to hold their faces still.

At the same time, conventions from portraiture, an expensive, serious affair bereft of smiles, shaped people's postures in front of the camera. Portraiture, which was the custom and form that preceded photography, tended to banish smiles. As Mark Twain is supposed to have quipped, "A photograph is a most important document, and there is nothing more damning to go down to posterity than a silly, foolish smile caught and fixed forever."[8] At the first photo studio in London, established in 1841, daguerreotypist Richard Beard told sitters to say "prune" to purse their lips shut. While photography interacted with local customs and artistic traditions in different ways across the globe, its arrival often imposed new rituals of self-presentation. Japanese photographer Matsuzaki Shinji's *The Essentials of Photography: Dos and Don'ts for the Photographic Customer* (1886) enumerated twenty-six social protocols and procedures for sitters to prepare for a studio session.[9] "Once you have decided to have your photograph taken," he wrote, "you should clean your entire person, comb your hair, shave your face (while those with long beards should wash them thoroughly), and take care that no dirt is attached to the face or the rest of your body."[10] A serious business indeed.

The gaze of the camera has always exerted an arresting influence on the bodies of its subjects. The lengthy exposures needed by early photography cultivated poses that were easier to hold than a prolonged smile. By the 1850s, it was possible in the right conditions to take photos with only a few seconds of exposure time. As camera technologies quickly matured, that duration became shorter and shorter, eventually contracting to the almost instantaneous snap we know today— once known as the "Kodak moment." By 1900, Kodak was making cameras with a shutter speed of 1/100th of a second. Even so, photographed smiles were rare until a few decades into the twentieth century. Even now, when our exposure times are faster than the blink of an eye, the mere lens of a camera—or a phone—is enough to prompt a struck pose, an automated rictus creasing our faces. Like deer in the headlights, we stop and tease our faces into their camera-ready mode, pulled out of time for the briefest of instants before returning to our regularly scheduled programming. We are habituated to the camera's stare.

Inventor George Eastman launched the first successful roll-film hand camera, the Kodak, in the summer of 1888. A simple box camera preloaded with a 100-exposure roll of film, the Kodak made photography accessible to amateurs without artistic or technical training or expertise. Once the film roll was full, these amateur photographers would send the entire camera back to the Kodak factory in Rochester, New York. There, the company would reload the camera with a blank roll and return it to the customer while developing their film. As Eastman's ingenious marketing campaigns crowed, showing women and children joyfully operating their cameras with ease, "You Press the Button, We Do the Rest." Snapshot photography was born—and it was fun and simple enough for a child (and a woman, will wonders never cease!) to use.

Eastman had turned photography into a consumer enterprise. Within the decade, the Kodak unleashed over a million amateur shutterbugs, who formed hobby clubs and launched magazines, and secured its place in history. Kodak entered the popular lexicon—"kodaking," "kodakers," "kodakery"—becoming a verb and a noun that transcended the brand. With Kodak's production of the one-dollar Brownie camera in 1900, which featured a removable film container, the company inaugurated the mass market for photography. The rise of amateur photography, enabled by the personal camera and commercially available darkroom development services launched in 1988, began to transform not only those behind the camera, but those in front of it.

As Kodak cameras continued to make their way into consumer hands, snapshot photography remade portraiture into a cheaper, more relaxed affair. The Eastman Corporation marketed the Kodak camera as a tool of consumer happiness, depicting a world thrilled by the technology in prominent national magazines across different markets, including *Harper's Bazaar*, *Cosmopolitan*, *Scientific American*, and *National Geographic*. By the 1930s, at least in much of the West, smiles started to be more commonplace, preserved for posterity. Some of the more famous ones include a photo series of smirking outlaws Bonnie and Clyde, discovered at a crime scene in 1933.

Like millions who came of age in the Kodak era, I grew up saying cheese.

In the 1940s, "say cheese" and its equivalents became de rigueur in consumer photography. The phrase's international cousins reveal the scope of the camera's powerful gaze. A family of words in different languages also tease mouths into smiles: in France, *ouistiti*, meaning "little monkey"; in China, *qiézǐ*, meaning "eggplant"; in Arabic, *yassar*, meaning" ease"; in Korea, *kimchi*. Yes, a lot of these words are food, but then food brings a smile to the lips, and a smile is global cuisine.

Of course, not everyone plays along. Some children—you know who you are—have a strong allergy to smiling in photos. My nephew, like my grimacing brother in his youth, makes a monstrous contortion of his face as soon as the camera turns on him. My niece switches on a megawatt smile, her little hands framing her face like the petals of a flower. My toddler affects a comical pout (I swear he doesn't get it from me). My dad wipes the smile off his face when the camera turns on him; the photographer who has documented several of my family's weddings tells us his secret is to shoot from his chest, where his camera seems to lie dormant. Unruly rebellions to the camera are routine, yet even these routines disclose the power of its intrusive gaze. When cameras look at us, they arrange our faces and bodies in myriad ways: they contort mouths into smiles as well as frowns, unfold fingers into peace signs, twist hips, and cock legs. Photography is a vision machine that, since its beginnings in long-exposure daguerreotypes, has demanded our bodies respond to its stare.

Kodak also catalyzed the next stage of transformation,

bringing digital immediacy to photography. In 1975, Kodak engineer Steve Sasson created the digital camera, going on to build the first DSLR (digital single lens reflex) camera with a colleague in 1989. Digital photography contracts the time of the photograph even more, condensing the camera and the darkroom into one device. Although Polaroids did this in 1948, the digital camera has allowed cameras to become mirrors, merging videos and photos. Although digital cameras didn't gain much cultural traction until the turn of the twenty-first century, they soon brought the electric eye to nearly every corner of human life.

Now, armed with powerful portable computers and increasingly impressive cameras embedded in them—mine boasts three lenses on the back, as well as a LiDAR depth scanner—smartphone users have developed a new, instantly recognizable repertoire of poses and a hyperawareness of flattering angles that were once the domain of professional models. And then, of course, there's the selfie. While endemic to smartphone culture, selfies are not a born-digital genre of photography. Selfies have been a native genre of photography since its very inception, with the first known selfie by chemist Robert Cornelius in 1839. Even so, selfies have achieved a peculiar dominance in everyday digital photography. Most smartphones now have cameras on the front and back, making it easier to quickly capture oneself in the frame. Digital natives—I'm looking at you, Gen Z—may not understand how big of a deal that is. But as someone from a cusp generation, I remember the clumsy acrobatics of rotating compact cameras to point and shoot ourselves, gambles that only paid off in flattering images some of the time. Our cameras have become mirrors, tools not just for seeing the world, but for seeing ourselves—sometimes to the exclusion of the world around us. Truly, the selfie has evolved into a genre all its own, one that focuses the camera's gaze like a magnifying glass heaving a sunbeam on an ant. In a selfie, everyone else disappears. Selfie sticks, a symptom and scourge of contemporary culture, best reveal this dynamic. In museums, oblivious selfie takers have backed into and toppled artworks. At concerts and sporting events, they have obscured others' view, or jabbed neighboring fans. At Disneyland, they were a menace on rides and a disruption to the carefully curated magic of visitor experience. Unsurprisingly, selfie sticks are now banned in many public spaces around the world.

The camera arrests with the power of its look. It coaxes smiles, but also disciplines. Used for surveillance, the same gaze that elicits smiles also turns the camera into a police officer. This dual function is best illustrated by an architectural embodiment of the power of looking: the panopticon. In the eighteenth century, the philosopher Jeremy Bentham conceived of an institutional building that functioned as a social control mechanism. As its name suggests, the panopticon is built on the omnipresence of the gaze. A panopticon invests authority in a central tower, in which a watchman perches. All cells surround this tower in a circle, such that the watchman in the tower can see every cell, while prisoners in the cells cannot see the watchman. Knowing that they might be watched at any time, the prisoner internalizes that authority.

French philosopher Michel Foucault understood the panopticon as an important architectural metaphor for how modern power works. Signifying "the policeman within," the panopticon arrests a body in space, making people behave because of the mere possibility that they are being watched. Foucault's analysis of the panopticon is commonly used to understand principles and dynamics of surveillance in the world we live in. The panopticon can be understood as an ancestor of the all-seeing electric eye—raged against decades ago by Judas Priest and now embedded in far more pervasive, insidious ways. The consumer embrace of cameras throughout the twentieth century tilled the cultural soil for the world that ogles us from every corner today. Rather than just stopping to pose, perhaps we should have stopped to wonder. Consider how closed-circuit television has metastasized into surveillance fetishism, with consumers cheerfully buying

Rings, Google Nests, and other consumer cameras for their own doorways. Footage from these cameras is a fixture on neighborhood apps like Nextdoor, where neighbors ask one another to identify "porch pirates" who make off with their Amazon packages. Surveillance is now a consumer enterprise, although well networked into policing systems. In California, police have deputized Teslas that may have been parked in proximity to crime scenes, in some cases obtaining warrants to access data they may have captured. Decked out with cameras and sensors, these vehicles are treasure troves of environmental data. Cameras are everywhere, and we don't know if and which human eyes are behind them. But we behave as if someone is watching. And that, of course, is the point.

This era of surveillance may seem a more blithely dystopian exercise of the camera's look than having to say "cheese." Yet its similar effects on the observed builds on the arresting power of the camera's gaze. Cameras are conscripted in service of policing our bodies. In very real ways, they determine how and where we can move, where we can or cannot go. In airports, X-rays visually unpack our bags, allowing security workers to peer into our belongings. Flagged, we are pulled out of line to watch a uniformed stranger rifle through our possessions. Backscatter X-rays, another airport imaging tool, undress us as we stand with legs spread and arms raised above our heads. Facial recognition, increasingly used in official areas, and at the time of writing is in the process of being integrated into airports, determines if we can pass and where we can go. As early photography did, these electric eyes stop us in our tracks and insist on looking.

More alarmingly, this has become an entire ecosystem of vision machines. We've outsourced looking from our bodies so much that now we're being looked at by nobody in particular; instead, electric eyes—the smart cameras that watch our babies, monitor our homes, sentinel our streets, and guard our airports—have real impact on the bodies they observe.

In late 2022, attorney Kelly Conlon chaperoned her daughter's Scout troop to storied New York City venue Radio City Music Hall to see the Rockettes. Shortly after passing through the metal detector, Conlon was pulled aside and asked to confirm her identity. She was then denied entry. Facial recognition software had identified her as an employee of a law firm involved in litigation against Madison Square Garden Entertainment, Radio City Music Hall's parent company. The company had placed her, along with other enemies, on an "attorney exclusion list," enforced with the assistance of software that matched unwelcome interlopers against profile photos on their firms' websites. This incident, a terrifyingly petty application of facial recognition by a corporation, highlights the sinister dynamics of machine-to-machine seeing. Unchecked, machines do the looking, with humans, in this case security, merely enforcing their judgments. Other private entities have used facial recognition in suspicious ways. In China, some hotels in major cities have been scanning guests' faces at check-in, leading the Chinese government to ban excessive use of the technology by private enterprises.

As the camera's gaze has permeated public space, it has recast our bodies as data. Facial recognition reduces human faces to information. It remakes faces from the living, evolving landscapes of personal history and memory that we know to a topography of data—rethinking them in terms of activations, key points, eigenfaces, classifiers, and polygons. Facial recognition software parses a face's unique geometry, extracting and analyzing features from images by measuring the distances between points. So parsed by computer vision, a faceprint is less a gestalt of eyes, nose, mouth, chin, and brows than it is a calculated map. We are reduced to geometry. As the Jacquard loom turned the knowledge and intelligence of our hands into information, the camera's stare turns our physical features, faces, and movements into data. Pared to our facial geometries, our data doubles—informational twins of how we look, move, and behave, not to mention what we buy and consume—become the metric against which we are measured.

We are surrounded by electric eyes. Facial recognition is one feature of an era defined by the widespread automation of vision. Our vision machines are evolving human eyes to the margins of looking. Photographs were once always taken with human eyes as their ultimate destination. The twenty-first century has seen the advent and application of computer vision and machine learning to most images in circulation; most existing images—far more than the human eye can see—are generated by and headed for machines. Whereas surveillance in the era of avian reconnaissance photography (hot-air balloons, pigeons, planes) was destined for human eyes, now it is machine-generated imagery that is also machine-readable. And through this economy of machine-readable imagery, decisions are made about how other bodies get to live and move. As artist and geographer Trevor Paglen describes in this era of machine-to-machine seeing, "Your pictures are looking at you."[11] We inherited our well-behaved stillness from photography's fledgling moments. Those early photographers who made the camera glare at sitters were already rehearsing this relationship. With its capacity to freeze living bodies into a quiescence resembling death, early photography foreshadowed its logical conclusion in digital vision machines that wring the life from our very movement.

In a mall a ten-minute walk from my childhood home,

there was a FUJIFILM kiosk where I regularly took rolls of film to be developed. While I had been the photographer for most of the pictures on the roll, carefully squinting through the viewfinder as I pointed my compact camera at my family or friends on field trips, I always felt a tickle of excitement when I opened the envelope of photographs. I knew what to expect, but there were always surprises: an unwitting passerby, someone's eyes closed at exactly the wrong moment, or the specific contours of my little brother's cross-eyed monkey face. Analog photography, that is, photography that requires film to be developed, needs human eyes to be seen. Otherwise, images lie dormant. That selfie you just took, by contrast, is machine-readable whether or not you summon it to your screen from your bloated personal archive. From pixel data to geolocation, digital photographs are embedded with reams of data about how we look, where we've been, and increasingly, how we might live in the world to come.

Every two minutes, we collectively take more pictures than were taken in the entire nineteenth century. Each year, we snap an estimated 1.8 trillion photos, many of which we will never look at again. On Facebook alone, 300 million photos are uploaded every day. In the era of social media, when many of us feed images of ourselves and others into corporate databases, this information becomes chum for data-hungry algorithms. Our faces—machine-readable, tagged, and time-stamped—have become training data for artificial intelligence algorithms that are being leveraged in increasingly unsettling ways. Controversial facial recognition company Clearview AI has, without consent or regard for privacy, scraped billions of facial images from across the Internet and social media to construct a database so vast and invasive that international watchdogs have called it illegal. These images—ours—are sold to governments, to police, to anyone who pays. In many ways, we've relinquished seeing to an invisible, planetary field of electric eyes that transform light and flesh into data points. From automated license plate readers that check if we've paid our tolls (and ensure we're billed if we haven't) to insidious facial recognition tools that quietly hike insurance rates or deny access to services, these vision machines regulate how we live and move. Often deployed by governments and corporations, they operate in service of the marketplace.

And what, then, of human eyes?

Even as photographic and imaging technologies have expanded the field of human vision far beyond the borders of the skin, that scale has come to dwarf human concerns. As Trevor Paglen suggests, "If we want to understand the invisible world of machine-machine visual culture, we need to unlearn how to see like humans."[12] At the same time, we can no longer trust our eyes to verify the images that we do see.

Instagram filters that smooth over facial imperfections are a dime a dozen, making it difficult to know what someone looks like in meatspace. Photoshop, once revolutionary, now feels quaint compared to AI-powered image generators that can fabricate entire scenes in seconds, further reconfiguring human vision. Whereas Photoshop altered the real, text-to-image generation tools such as Dall-E and Midjourney conjure it from scratch, accelerating the erosion of our trust in our own senses. While images generated in the emerging moments of the technology had several tells, notably weird hydra hands with too many fingers in the wrong places, AI-generated bodies now are increasingly polished, though still shaped by the racist, sexist, and ableist biases of the datasets they draw from. The deluge of "AI slop," as it has come to be called, overrunning the Internet, testifies to the domination of the machine-to-machine visual economy. Even when we are looking, we aren't certain what we see. In an era of electric eyes, what value does human vision still hold?

Paglen reminds us that the visual strategies honed in human-human culture may falter in the face of these new machine-machine systems. Still, people try. Some opt for camouflage: styling hair and makeup in ways that foil pattern recognition, growing facial hair, or wearing masks. Others don anti-surveillance fashion, which includes clothing with reflective stripes, glitchy prints that confuse algorithms, and sunglasses and fabric that block thermal detection. And some resist passively, refusing to participate at all. I often see parents on social media blanking out their children's faces with icons—if they post them at all.

The history of human-machine looking tells us we are too easily coaxed into standing still—for posterity, for surveillance. Cameras have trained us to pose, to submit to their gaze. But like all cultural habits, this one can be questioned, and perhaps unlearned. Interwoven with operations of the marketplace and the state, vision machines prioritize efficiency and conformity. If there is hope, it lies in our very human capacities for inefficiency, for unpredictability, for care. We can outsource looking to machines, but we can't outsource empathy, self-expression, decency, or wonder.

Much of the current criticism of AI-generated art, which has come a long way since Obvious's lousy *Portrait of Edmond Belamy*, orbits around questions of authorship, authenticity, and the threat to human labor. These debates echo the early skepticism that met photography in the nineteenth century. French poet Charles Baudelaire, for instance, famously denounced photography as the "mortal enemy" of art, fearing it would corrupt the imagination by anchoring it in a soulless, mechanical process.[13] But photography evolved. It became not only a tool of record but also a medium of creative expression, a way of seeing and sculpting the world.

And as its aesthetic meanings have shifted, so too has its relationship to human looking. The meanings of these technologies are never ossified. Even as photography, and now computer vision, threaten to pin us in place, to render us sortable and knowable, we don't have to remain still. We can look again. Differently. Remember, *Earthrise* moved us not because it was captured, but because it was witnessed.

5

root

I USED TO LIVE IN A SHARED HOUSE IN SYDNEY, A SHAM-
bolic Victorian with a wrought iron balcony and dark green
windows. My housemates and I were all living the spacious,
exploratory lives of young adulthood, rich with time and
hungry for experience. It was one of the best times of my life,
each day wide open to serendipity. One evening, one of my
housemates brought a fresh love interest home to plot their
evening's adventure. Over cartons of Thai food, they shared
their plan. They were off to play Green Man.

Green Man, as I learned that night, is a game that in-
jects chance into the urban landscape, like a metropolitan I
Ching divining meaning from randomness. Players embark
on a walk, and whenever they reach a traffic light, they cross
wherever the green man beckons. Green Man rejects urban

purpose and defies destination. The game might lead players in circles or it might surf them somewhere unexpected, depositing them in new locales—an unfamiliar bar, an unseen mural, or a park unknown. Either way, it requires a paradoxical embrace and rejection of the urban environment. Green Man invites players to redefine cities as exploratory landscapes, ripe for discovery. And unlike the roads that channel people to their destinations—the commute, say, from home to work—it turns those conduits into avenues for fun and play. At once silly and serious, Green Man embodies the power of walking in cities, and of the primal force of the foot as a creator and explorer of the built environment. This chapter descends from the planetary perspective to the ground, where walking invites a slower, more embodied way of knowing.

According to Aaron Sussman and Ruth Goode in their guide to the "magic of walking," "the body is built poorly for sitting, only a little better for standing, but it is unrivaled for walking."[1] The human foot is a marvel of evolutionary engineering that distinguishes us from other animals. The first hominins, the earliest members of our lineage, didn't have large brains like modern humans, didn't use sophisticated technology, and didn't talk. They did, however, walk on two legs. Our feet are the very foundation of modern humanity as we know it. Bipedalism is the most ancient human adaptation, setting the stage for everything—our reliance on tools and technology, language, dietary flexibility—that marks us as human. It freed human hands for tools and communication, and breath for speech. Walking—upright, that is—is as central to our humanity as writing and singing to one another.

Our feet embody this extraordinary legacy and history. As Leonardo da Vinci is said to have remarked, "The human foot is a masterpiece of engineering and a work of art."[2] Bearing our entire weight, our feet contain about a quarter of the bones and about a fifth of the joints in the human body. The humble human foot, unique and structurally distinct from every other paw, hoof, and foot on our planet, has not only powered our movement, but also enabled our subsequent cultural, political, and scientific evolution. In many ways, our feet have been as instrumental as our hands in coming to know and shape our world, albeit through a different kind of interface with the environment. We experience the ground through the soles of our feet. As that world has changed, with dirt paths reified in concrete, jagged terrain remade for human use, our feet have changed with it. One of our ancient technologies, shoes mediate a powerful relationship between foot, body, and world. Proponents of the barefoot lifestyle contend that "barefooting" is a gentler way of life, fostering attunement to the world around us by allowing ourselves to feel the land beneath us.

Walking is much more than the rhythmic transfer of weight from foot to foot, a dance of balance between feet, knees, and hips, or the most basic means of transportation. From promenade to protest, pilgrimage to parade, walks stage our encounters with the world. Walking, according to writer Rebecca Solnit, "is how the body measures itself against the earth."[3] Before the introduction of international standards, the human body was often measured against the human body; ancient cultures from Rome to China used some form of the foot to gauge length. Travelling by foot brings our entire bodies into relation with our environment and with one another. It is an expression of the interconnectedness of body, brain, and world.

An engine of the mind, walking has long captivated philosophers. Some of history's finest thinkers were also ardent walkers. Enlightenment thinker Immanuel Kant was known by his fellow townspeople as the "Königsberg clock"; so regular was his 3:00 p.m. walk that you could tell the time by his figure ambling past. Political philosopher Jean-Jacques Rousseau characterized his legs as the motor for his mind: "I can only meditate when I am walking. When I stop, I cease to think; my mind only works with my legs."[4] Existentialist philosopher Søren Kierkegaard confessed, "I have walked myself into my best thoughts and I know of no thought so burdensome that one cannot walk away from it."[5] Friedrich Nietzsche rediscovered himself through epic walks, eventually aligning the movement of walking with the stirring of spirit. Walking—not running—is defined by these philosophers, as well as myriad others who hike and wander, by ease. When you hit your stride, foot, body, and mind are in communion and communication with the landscape.

As our hands make marks on paper through writing or drawing, our feet carve paths into the earth, whether grass, soil, rock, or clay. Our feet, like our hands, make worlds. In his poem, "Writing with One's Foot," Nietzsche imbues the foot with this creative power:

"I do not write with hand alone:

My foot does writing of its own.

Firm, free, and bold my feet engage

In running over field and page."[6]

Where our intelligent hands hold knowledge in their palms, our feet know where we've been and lead us to where we are going. Walking is a conversation and collaboration of body, mind, and place.

Paths trace this dialogue between foot, body, and environment. The desire path, my favorite kind of passage, reveals the terraforming power of our feet. The desire path is the unsanctioned dirt way trodden into grass, and the shortcut through the park. It is the dusty track scooting under a broken wire fence, and the etched dirt veering at a jaunty angle from the pavement, an unruly alternative to concrete. Worn into the dirt, these paths register the desire of the many footsteps that have carved it into existence. So understood, routes at all scales might convey human desire—the desire to go, to grow, to connect, and to settle. On a larger scale, walking has forged roads and trade routes. It has sculpted the contours of parks, towns, and cities. And on these ways, the history of human movement reverberates. "Always, everywhere, people have walked, veining the earth with paths visible and invisible, symmetrical or meandering," writes Thomas Clark in his enduring prose-poem "In Praise of Walking."[7]

In *The Old Ways*, British writer Robert Macfarlane explores ancient routes that interlace the British Isles—across land, water, and beyond. He hikes such venerable tracks as the Ridgeway, used continuously by British travelers for at least five thousand years, and the Broomway, a perilous route over the English foreshore that is only accessible at low tide. In so doing, he reads the history of the land and its people, etched into the landscape. Macfarlane describes paths as "the habits of a landscape" and "acts of consensual making."[8] More than a consensus between people, these ways evolve from the collusion of moving bodies and the environment. The first paths may have emerged organically from animals and humans traversing territory—all paths are desire paths, really—but as they evolved into roads, they began to shape human migration.

Ways are an ever-present, ongoing dialogue between foot and world, living relationships that need to be nourished in even the most developed of infrastructure. The most primitive are magnetic egresses, the paths of least resistance or greatest compulsion. Abetted by technologies that turn paths into roads, they are sites of human progress. Cities arrived on foot, first as settlements, then villages, then towns. Ancient paths blossomed into roads, scaffolding corridors for trade, communication, and migration between settlements. In the ancient Roman Empire, cities erupted along the network of roads that connected the Roman world, routes often with humble beginnings.

Some contemporary urban infrastructure springs from these prehistoric ways. Construction on Rome's oldest road, Via Appia Antica, began in 312 BCE. Many roads in London are of Roman origin, with a two-thousand-year vintage: Old Street was already called Ealdestrate in 1200 and is now a favored canvas of street artists such as Banksy. Liulichang Street in Beijing was a nexus for scholars, painters, and calligraphers to gather seven hundred years ago, and today still offers a respite from the bustling city for those seeking art supplies, musical instruments, and jade carvings. Al-Mu'izz

Street, a major avenue in the walled city of historic Cairo, was one of its most important arteries, home to numerous souks and monuments that are a testament to Egypt's history. The Grand Trunk Road, extending from Bangladesh to Pakistan through India, built in the third century along an already ancient route and still very much alive today, has morphed into a series of interconnected national highways. In this very real way, walking constitutes place.

Roads and sidewalks are, of course, technology, although we don't tend to think of them in that way. Like all technologies, they embody invisible values about who and what can, and should, use them. They are, as Ursula Le Guin reminds us of technology, part of the "active human interface with the material world."[9] Roads are concrete reckonings with our environments, making travel and trade possible. More than that, they channel how we move, and so steer how we live. All roads are navigations between freedom and discipline— and as our cities become increasingly digital, this dialogue becomes increasingly fraught.

You open your front door and descend the stairs of your apartment block, which deposits you through a lobby onto the bustling streets below. You turn right and wait at the crosswalk, until the green man jauntily signals your passage. According to modernist architect Le Corbusier, "a house is a machine for living in."[10] The same can be said for cities. Cities are many things: settlements, spaces, and systems. The built environment might also be understood as a movement technology, funneling the bodies of its citizens along channels sculpted by urban planners, as well as along the roads chiseled by the footsteps of history.

Walking composes the city, even as the city composes the capacity to walk. Comprising interlaced roads both rigid and meandering, cities are formed of footpaths, but they also choreograph footsteps. Embellished by infrastructure technologies—traffic lights, speed bumps, and street signs, as well as telephone poles and communication lines, electricity cables and boxes, and sewers—cities make choices about where and how bodies should go. In many ways, urban design and urban planning are as much about choreographing movement as they are about creating human dwellings. Some cities prioritize machines over people, relinquishing precious walking space to cars and parking spots. Whereas the previous chapter, "Eye," explores how technologies choreograph stillness, this one considers how they orchestrate movement.

How people walk—at a leisurely amble, a purposeful power walk, a playful strut—not only reveals their way of living and being in the world, but also shapes their own experience of it. In cities, walking is one interface with the built environment. From the rudimentary paths that walking wears into the world to the more engineered lanes, roads, and highways in cities, the built environment channels our bodies in particular ways. Traffic lights usher us along at regular intervals, soundtracked by beeps of varying urgency. Under- and overpasses shunt humans out of the way of traffic. Public transportation maneuvers people into buses and trains, hurtling them to schools, office buildings, and institutions. Sidewalks guide us to certain destinations. If we walk mindlessly, as on a commute, a city automates us.

Like the Green Man game my friends played on their date in Sydney, through walking, people have long sought to inject fun and freedom into urban environments. By walking playfully, these activities reject the automatic choreography of the city. In this respect, this game is one version of an enduring creative strategy in modern cities. Cities might automate our feet and compose individual people into masses, but they might also be stages for individual expression.

The urban stroller is a common figure in stories of freedom in the modern city. From Leopold Bloom in James Joyce's *Ulysses* to Virginia Woolf's *Mrs Dalloway*, wanderers have written their own stories as they navigated cities on foot. Charles Baudelaire named such a figure the "flâneur," an urban explorer and observer of modern life, who originated as a literary type in nineteenth-century Paris. This person, often an independently wealthy dilettante, was able to wander the city at their leisure rather than hurry along to work or school. Freed from the commute and the need to earn, the flâneur's eyes, powered by their feet, offered profound insight into the cities that were then redefining modern life. Guided by the environment—alive, constantly changing, a cauldron of flesh, concrete, glass, and machine—rather than a destination or the legislated channels, these renegade strollers offered new insights on the city and asserted individualism by merely walking. The flâneur found new forms globally, inspiring generations of strollers looking to imprint their own footprints on the metropolis.

The modern city was the crystallization of advanced capitalism; its architecture presses hands, feet, and bodies in service of labor and consumption, even as it alienates people from one another. The Situationist International, a European avant-garde collective of social revolutionaries comprising intellectuals, artists, and political theorists, understood cities as engines of human automation, dulling its citizens into numb lives of malaise and boredom. By this account, most people are sleepwalking through metropolitan life. The Situationists offered walking as one counterstrategy to this urban stupor: the *dérive*, or drift, as the "technique of locomotion without a goal."[11] Emphasizing arbitrary routes—say, following the green man's signal—the dérive is a subversive tactic meant to reclaim the streets for play. Such, after all, is the power of walking.

In the nineteenth century, the flâneur bore witness to a series of seismic transformations taking place in the city. Walking was still a primary mode of getting around, alongside vehicles such as horse carriages, carts, and wagons. Subways and metro lines, beginning with the 1863 opening of the Metropolitan Railway, now part of the London Underground, expanded the city and made it possible for people to quickly reach distant areas. Streetcars extended the foot's range on the city surface; the 1881 debut of the world's first electrically operated streetcar in Berlin made a profound impact on the range of the human foot. These transit technologies, transformative as they are, and as much as they bent time and space to augment the reach of individuals, were still public, constituting public spaces within and beyond their networks. Walking cities, and later, streetcar cities, tended to concentrate social, cultural, and business life at its core. As infrastructure, public transportation composed time and space in ways that cultivated collective living, mobilizing citizens into throngs. They also carved out undesignated niches of time for riders to spend as they will. This time has also fostered a rich culture of expression around public transit. Many commuters have finished reading, even writing, novels on commutes, certainly a much riskier endeavor while driving. In these suspended pockets of movement and waiting, new habits and forms of attention could take root. Joel

Sloman's *Bus Poems*, for example, contains verses written to enjoy for the duration of a bus ride.

The advent of the car in the twentieth century transformed the landscape of many cities. Highways were introduced to existing cities, sometimes in ways that preserved their walkability and sometimes otherwise. European cities often used ring roads to bring highway traffic to the city, while maintaining the historic human scale and walkability at the city core. Many American city centers were gutted by roads and highways, becoming less hospitable to feet and exiling walkers to sidewalks—if they bothered to build them. Transformed by car culture, not to mention automobile lobbies and companies, such cities relinquish public space to highways and, with it, communities once networked by foot. With the dispersal of populations into suburbs, social and cultural life is also more diffuse. Encasing and isolating drivers in hard shells, cars privatize and destroy public space. Automobile infrastructure—roads, highways, thronged intersections—creates distances only accessible by car, at the cost of other forms of transportation and walkability. Cities defined by car culture, famously sprawling metropolises such as Los Angeles, are far less walkable, particularly without corresponding efforts to develop public transportation infrastructure.

In many cities, public space is being eroded. Malls and

parking lots define the look and feel of numerous contemporary cities. Remote technologies that don't require people to leave home have contributed to the demise of stores and services. Coupled with the impact of cars, urban strolling, according to philosopher Frédéric Gros, has become "more difficult, less delightful and surprising."[12]

The demise of public space has inspired imagined visions of alternative futures that prioritize the place of our feet in public infrastructure. In the postapocalyptic novel *The Fifth Sacred Thing*, Starhawk imagines cities remade by disaster. Rather than the dystopia that usually follows apocalypse in popular representations—zombies, murder, deserts, devastation—in this tale San Francisco has been remade as an ecotopia. When fuel begins to dry up for everyone but the wealthiest, citizens tear up the streets with pickaxes and fill them with compost. Roads, once beholden to cars, give way to garden promenades restored for walking.

One of my favorite murals in San Francisco is Mona Caron's *Market Street Railway Mural*, which envisions something like this ecotopian future. The mural depicts a bird's-eye view of Market Street, a major arterial in San Francisco that runs through the Castro, then Downtown, and all the way to the Bay. The image begins at the Ferry Building, where Market Street originates, in 1920, lined with train tracks, streetcars, as well as passenger automobiles.

Flocks of pedestrians line both sides of the street, dressed in era-appropriate wool suits and hats. The next section of the mural depicts Bloody Thursday during the longshoremen's waterfront strike in 1934, with union workers and policemen in bold clashes among clouds of tear gas. The next section is a rendition of a Labor Day parade on Market Street in the late 1930s or '40s. The mural continues to transform through time and space: a 1980 Gay Pride parade, daily life circa 2000, a 2003 march against the Iraq War—eventually crescendoing in a bright and speculative future in which light-rail, bicycles, cyclists, and pedestrians (as well as a mahout riding an elephant and people astride a pair of ostriches) cohabit in vibrant peace among islands of trees, grass, and shrubbery. It's an exuberant portrait of a city reimagined—not for traffic, but for life. This mural depicts how formative our feet have been, and could, be for our metropolitan landscapes.

While San Francisco hasn't quite ripped up its roads or turned them over to ostrich steeds, its streets are something of a laboratory for the future of the human body. San Francisco residents recently voted to permanently transform a stretch of road known as the Great Highway into a park, a movement initiated by pandemic-era closures intended to create more outdoor space for residents to walk and bike while social distancing. There have been a number of efforts globally to remake urban public spaces by reclaiming city streets for pedestrians and cyclists. In the 1970s, the Dutch city of Groningen excavated the highways in its city center, expelling cars from this area and exiling them to an inner-city ring road. In the decades hence, the city has become a haven for cyclists and pedestrians. In 1999, the once polluted and traffic-choked city of Pontevedra, Spain, launched an effort to expel cars from its center, pedestrianizing over three hundred thousand square meters of the city. In the 2000s, the city of Seoul, South Korea, razed an elevated eighteen-lane highway that covered the Cheonggyecheon Stream. The resulting pedestrianized waterfront has revitalized downtown.

These examples show how walking can become a regenerative force in urban life. Re-centering the foot—and the body more broadly—transforms how cities feel, function, and flourish. Designing cities for bodies instead of machines is not simply aesthetic, but ecological. Walking becomes a way of reclaiming place, rhythm, and relation—a step toward more livable futures.

As vehicles for our bodies, our feet are a primary interface between self and world. With the advent of self-tracking technologies that turn our footsteps into information, they, too, have become fodder for systems that flatten the nuance of lived experience. That one must walk a daily ten thousand steps for health has become almost as much of a maxim as that ancient adage, "A journey of a thousand miles begins with a single step." It might be truer to say instead that a journey of ten thousand steps begins with a single pedometer.

The recommendation originates in a marketing campaign by a Japanese company, the YAMASA Corporation, which had introduced the world's first wearable pedometer during the 1964 Olympics, a device they called a *manpo-kei*, or "ten thousand step meter." The number was chosen arbitrarily, based on the Yamasa's design team's hunch that it felt right. In the decades since, this figure has become a common metric for health. The ten-thousand-step challenge is a website, an app, and to some, a competition. For a time, my brothers were regularly comparing step counters to see who more regularly hit the mark. Despite its popularity and absorption into our popular health lexicon, ten thousand steps a day isn't as meaningful a yardstick as these companies would have you believe. Although health agencies such as the World Health Organization and the Center for Disease Control and Prevention have adopted the recommendation at various times, the science behind it is dubious and contested.

Admit it. You have checked your daily step count to see how you've fared. I'm a bit lackadaisical about it myself, only checking my Health app on days when I suspect I've crushed it, far surpassing the ordinary circuit between my office chair and my kitchen counter. And it gives me that reliable hit of dopamine when my device verifies that I have indeed exceeded the recommendation, or on days I've gone on hikes and ascended over fifty floors. It's a telling feature of the technology that height is measured in floors, displacing the intuitive sense of feet on mountains with an architectural unit. On one daylong hike through Yosemite National Park, it informed me I had scaled eighty-seven floors—the height of the Lakhta Center, the tallest building in both Russia and Europe.

Whatever it's worth, ten thousand is a neat and sticky figure, a tempting and reachable goal for many whose lives are rife with dissatisfaction and short on time. For these reasons, ten thousand steps a day is an emblem of the quantified self. The "quantified self" names practices of self-tracking abetted by such devices as our smartphones, Fitbits, Apple Watches, and Oura rings. At its most idealistic (or delusional?), self-tracking is an assertion of ownership and agency over personal data by empowering "self-knowledge through numbers." By integrating data acquisition into everyday life— and being the arbiter of that data—quantified self practices enable people to improve their mental and physical performance. At their most extreme, quantified self practitioners are essentially optimizing themselves. But what are they optimizing for? Data fetishism reduces the rich complexity of our bodies and behaviors to information that lacks nuance. Self-tracking encompasses far more than footsteps, of course; the more sophisticated trackers gather data that includes heart rate, body temperature, and sleep habits. But our feet have been the gateway drug. Like the rest of our body, our footsteps have become data—and we measure ourselves against this information architecture. Increasingly, this information architecture is merging with the built architecture of cities. As it does, it displaces the ancient conversation between foot and path that has shaped all ways, and the towns and cities that emerged from them.

The emergence and dominance of dynamic mapping tools like Google Maps has also had a profound impact on how we move through space, as well as how we conceive of it. Pinned on a dynamic map by our smartphones, we engrave digital paths all around the city. It is an incredible boon to be able to chart, in a matter of seconds, a route to a place I've never been. We do not know these routes like the ancient ways, where history is worn into the land itself (although sometimes they coincide). Google Maps never charts a desire path; it only surveys the known world. Instead, a digital route computes the fastest path, the projected time it takes to walk based on the "average" human body.

Since my smartphone entered my life, I have found that it has taken me much longer to discover a place anew, to learn it by foot, to know it in my bones. My mind is in my device, my footsteps forecast by the path it legislates. I pay less attention

to the world around me, aligning my feet with the path optimized by the smart system that tells me it will take four,
fourteen, or twenty-five minutes to walk to my destination,
at the average walking pace, whatever that is. This channel
of information—from your feet to your phone to Google's
GPS system—is but one stream of the kind that feeds the
smart city. The smart city gorges on information gathered
from its denizens as they live, move, work, and play in it.
The quantified self merges with the quantified community. A
nineteenth-century flâneur would be frightened and dazzled
by the way these mysterious invisible currents shape one's
experience of the landscape.

Optimization through tracking is not only the province
of the individual, however, and is leveraged and abused by
entities such as governments and corporations. In one version of this, these kinds of systems turn us into digital cogs,
our footsteps conscripted in service of production. Amazon's
warehouse floor, for example, is governed by an exacting and
exhausting surveillance system that tracks how long workers take to move a package from order to fulfillment across
the massive space. Measured against punishing productivity
quotas, they are perpetually rushed, under-rested, and over-
worked. Designed for efficiency, such a system denies the id-
iosyncrasy of human movement and understands walking in
economic terms, not human ones.

Amazon's warehouse floor and contemporary cities may not be the same environments, but they share investments in surveillance and tracking infrastructures that aim for optimization. As the enduring popularity of the ten-thousand-step goal shows, digital health data can become less metrics and more goals—and in some cases, punishing targets that are simply not available to all bodies. Like the flâneur and the Situationist, luxuriant in their navigations through the city, inefficiency is one model of resistance and individuality.

Built environments are interfaces that organize bodies in space and time. While walking may be the most natural thing in the world, it is increasingly being integrated into technological systems. Impregnated with information-gathering sensors, smart cities are the inexorable conclusion of this logic. Cities are becoming algorithmic labs for human movement.

If, as Le Corbusier said, a house is a machine for living in, then is the smart city a computer? And if so, are our footsteps code or cogs? Various technology metaphors have been applied to cities, as thinkers have raided contemporary science for ways to think about the spaces in which they live. Cities have variously been described as animals, ecosystems, and machines. The city-as-machine has taken several forms, each echoing the dominant technologies of its time. The industrial era saw the factory as a common urban metaphor.

Continuing this thinking, the smart city has been likened to a computer. The city is also used as a metaphor for computers. In software visualization, cities may be used to represent complex systems as structures and buildings. Either way, computation and urban life are intimately entangled, and our movements become inputs in their shared system.

According to urban theorist Shannon Mattern, such metaphors matter because they condition urban design, planning, policy, and administration. When cities are conceived as computers, city planners are wont to treat citizens less as people and more as information. Organizations, from think tanks to the United Nations, have sought to standardize a set of benchmarks for determining the health of cities, ranging from energy use and average life expectancy to emissions and the homicide rate. As Rob Kitchin notes, these urban indicators "become normalized as a de facto civic epistemology through which a public administration is measured and performance is communicated."[13] The collection of such indicators inform a model of "dashboard governance" by which a city's leaders then make decisions about where to invest resources. And indeed, city dashboards, or as tech company Siemens once called theirs, the City Cockpit, have been implemented around the world, from Rio de Janeiro to Mangaluru, with varying success. Bloomberg New Economy maintains a Dynamic Cities Dashboard, which tracks eighteen metrics across thirty cities from fourteen geographical regions that represent six city archetypes defined by economy and population growth, including Hong Kong, Mexico City, Stockholm, and Istanbul. This data is organized around six pillars (Fair, Happy, Sustainable, Data-driven, Innovative, Responsive) against which the tool audits progress.

Surely, more information about our cities can't be a bad thing. After all, how are we to improve the quality of life for urban dwellers if we don't know anything about how much they earn, the rent they pay, the air they breathe, and the dangers they face? But if something can't be measured, then it does not count—and much is lost when reducing unruly bodies to mere data. This model of urban intelligence overlooks the vitality that makes cities truly sing. There are numerous ways of knowing and being in a city that can't simply be counted: the local, situational forms of intelligence and experience that reside in our very bodies and minds, that drive how we move through the built environment, and that vary greatly for each individual in a city. It is the wisdom of the shortcut. The rhythm of the streets when the bars empty out. How sunlight spills down an avenue on a certain week in September. The unmarked entrance to the subway. The underpass where murals shift like seasons. It flows through the parent co-op and the nannies at the playground. Through the men who throw dice or play chess under the banyan trees at the local park on Sunday mornings. Through the small communities—the neighbors, friends, rivals, roommates, families, and strangers—gathered by affinity, habit, or urban necessity. A city's lifeblood, these ways of knowing and being, resist reduction. They shape how we dwell in a city, not how we are managed.

The advent of personal data collection technologies such as the pedometers built into our personal devices—smartphone, smartwatch, smart ring—makes it particularly tempting to conceive of the city as a computer. According to programmer Paul McFedries, there are two ways to think about this: "The city is a computer, the streetscape is the interface, you are the cursor, and your smartphone is the input device. This is the user-based, bottom-up version of the city-as-computer idea, but there's also a top-down version, which is systems-based. It looks at urban systems such as transit, garbage, and water and wonders whether the city could be more efficient and better organized if these systems were 'smart.'"[14] Yet, as artist Taeyoon Choi says, "If cities are computers for humans, they run busy software and often fail to compile."[15]

Groundbreaking on New York City's Hudson Yards began in 2012. An epic real estate development—the largest mixed-use private venture in American history—its five supertall skyscrapers house a high-end shopping mall, luxury condos, a school, and an office park that is home to several major companies. Beyond being a paean to corporations and the hyper wealthy, Hudson Yards was also conceived of as an urban lab. According to its developers, each day the sixty-five thousand people who traverse its residences, office towers, shops, restaurants, school, and open space would contribute to an enormous data stream that, when analyzed, would offer insight into the future of cities and how they should be built. A "data-rich research environment," as described by Hudson Yards' two developers and New York University, Hudson Yards was intended to siphon data about its users from where they go and how many steps they take to get there, to demographic information. It aims to "harness big data to innovate, optimize, enhance, and personalize the employee, resident and visitor experience."[16]

This view of the research platform is built on the assumption that these sensors, and the built environment in which they are ensconced, are but simple observers of human activity, rather than active shapers of it. Like all cities, how people use a space like Hudson Yards, and how they move through it, is choreographed by the architecture and space itself. These are not the ancient paths that blossomed from animal and human feet striking the earth.

A panoptic space embedded with sensors, Hudson Yards has not reached its lofty vision of a smart connected community. While it does gather data, it is not quite big enough to be especially useful as a research platform for cities. Moreover, even if the project was successful in gathering that data, it would tell very little about how people dwell in cities, or even in New York. It would essentially be information about how people move through Hudson Yards, or some optimized future version of it. What makes Hudson Yards so eerie is that it could be plopped in any global boomtown, as it is so devoid of Big Apple character and New Yorkers. If someone were interested in having the archetypal lifestyle of close-knit urban community televised on *Friends* or *Seinfeld*, Hudson Yards is not where they would find it. Nor would they find it nearby, as Hudson Yards is not built to integrate with, but to insulate from the rest of the city. This data isn't useful because it captures life in an artificial urban terrarium, not in the diverse, energetic, colorful scrum that shapes how urban bodies relate to New York. Living cities, like the roads from which they sprout, must develop in organic conversation between foot, body, and world.

The conceit of such a smart city model is that, through a raft of intelligent sensors, it would be in dynamic communication with its citizens, eventually optimizing the space for its citizens—and the citizens for its space. The metrics people choose to track in city dashboards—as with our own ten thousand footsteps—displace lived experience and become ends in themselves. Like the measurement and tracking of steps, coupled with other biometric information, behavioral and health data is used to optimize the self, and this information is also leveraged to optimize our cities, from how our bodies live to how our body politic lives. Ultimately, by integrating their movements into sensing apps on their devices, people become sensors themselves. In the smart city, surveillance, dataveillance, facial recognition, and the quantified self collide.

Cities have long shaped the foot, and they are products of the abiding dialogue between our footsteps and the environment. As streetcars and cars have reshaped our pathways and thus the feet that walk them, the infrastructural technologies of smart cities promise to reshape the interface of our bodies and environments. At the same time, they portend new norms of governance based on data and algorithms rather than the flow of human experience. Is the smart city a factory, cranking out robots, or can it instead recognize and honor the fluidity and diversity of human movement? In other words, what are the values of the smart city? Smart cities enshrine competing ideologies. On the one hand, they promise an optimized society built on human agency and improvement; on the other hand, they also control the bodies that dwell there.

Many cities have emerged from the encounter between

foot and land, visible in the worn paths that form their skeletons. But centering this history on the foot alone erases the extraordinary diversity of bodies that have inhabited, shaped, and moved through these spaces. Even today, much of the built world fails to cater to or express this diversity.

Although walking—randomly, playfully, purposefully, in lockstep—has always been as much a story of individuality as it is a story of conformity, it is not a movement available to all bodies. Despite the significance of walking in human history, not everyone can walk, after all. Not everyone walks in the same way or at the same pace, nor do they necessarily walk alone.

Cities—like all technologies—imagine and are built for an ideal user. Urban planning makes assumptions about how a "typical" urban citizen lives, works, and travels, as well as their needs, desires, and values. In traditional cities, this ideal user is an able-bodied young man; his stride is an easy match for a stair riser, the train schedule serves his nine-to-five schedule, he doesn't need a ramp or a place for his stroller. He stands comfortably at a phone booth, an awkward elevation, if not entirely out of reach, for wheelchair users.

In the 1970s, a group of young student activists at Berkeley, called the Rolling Quads, launched the modern disability movement in America. Frustrated with the inaccessibility of Berkeley's urban environment—their wheelchairs navigating an obstacle course of curbs, broken sidewalk, and sharp edges—the Rolling Quads staged a profound intervention into urban life. Apocryphal stories cast the Rolling Quads as midnight commandos, riding out with their attendants at night to smash curbs with sledgehammers and build their own ramps. The truth is a little less dramatic, though no less impactful, with the Rolling Quads pouring concrete to make a few renegade curb cuts along common routes. Fortified by a 1971 appearance at the Berkeley City Council, where they insisted that the city build curb cuts on every street corner, they issued a powerful call to action that launched the world's first widespread curb cut program. In 1972, the city of Berkeley installed its first official curb cut, eventually spawning hundreds of thousands across the United States.

While curb cuts were primarily installed for wheelchair users, they were soon found to be of great service to kids with bicycles, people with strollers, and anyone using wheeled devices who needed to cross the street. The curb cut effect refers to these unplanned benefits of designing for accessibility. Investing in one marginalized group could have unanticipated cascading effects for society. More walkable—and rideable—cities are essentially more livable.

The legacy of this advocacy offers critical insight into design justice, an approach that centers people who have historically been marginalized by design. As an approach, it understands design as a fundamentally political practice that makes choices about who it serves and who it leaves out. At its core, design justice is a theory and practice of participatory design, inviting and integrating communities most affected into the design process to help shape the spaces, tools, and systems that impact their lives.

One such example is DeafScape, an approach to urban design created by Deaf landscape designer Alexa Vaughn[17] that centers Deaf people and their communication and access needs. It is an urban application of DeafSpace, architecture that is designed to support the full expression of Deaf culture and experience, to landscape. DeafSpace emerged at Gallaudet University, the world's only university designed to be barrier free for Deaf and hard of hearing students. DeafSpace is engineered to support the richness of Deaf culture and communication through five units: Space and Proximity, Sensory Reach, Mobility and Proximity, Light and Color, and Acoustics and Electromagnetic Interference. For example, DeafSpace creates 360-degree sensory reach for visual and tactile information by designing view corridors essential for sign language communication.

DeafScape translates this ethos into the terrain of the city. Understanding landscape as a rich sensory experience, DeafScape is designed to amplify visual, tactile, and even olfactory cues for orientation in the absence of sound. Vaughn emphasizes wider pathways, so there is enough room for two or more people to communicate in sign language. More than just accommodation, DeafScape aims to integrate form and function; visual rhythms along sidewalk edges aid in spatial understanding, and textured transitions are indicators for the sidewalk, planting areas, and the street. Such textured features are already prevalent in many cities; some more common forms of urban design that incorporate sensory feedback to the feet are tactile paving, textured ground surfaces that can often be found at roadside, on stairs, and on railway platforms, to cue people with vision impairments to changes in the built environment.

DeafScape exemplifies the promise of purposefully multisensory design: an approach to design that recognizes and celebrates the plurality of embodied experience. It reimagines the built environment as an arena for human communication as well as human movement—something we would do well to remember when machine-to-machine communication plays an increasing role in choreographing human activity in cities. As smart cities have transformed urban paths and surfaces into information-gathering interfaces, how bodies move through the city becomes an interaction design problem as well as a human one. And it is not only a question of design, but of justice.

Automatic systems now choreograph moving bodies in increasingly explicit ways. Food delivery couriers and Amazon delivery drivers are sent around the city at the bidding of apps, their scooters, bikes, vans, and bodies frenetically texturing the cityscape. And just as the Ford Model T, the first affordable, mass-produced automobile, catalyzed the transformation of many cityscapes for wheels over feet, self-driving cars are poised to transform urban landscapes for walking once more. Waymo, a Google subsidiary for autonomous vehicles, has been testing its robotaxis in working cities since 2021. My city has been quite a test bed for these driverless cars. I don't walk five minutes without seeing one. At first, I was cautious, reluctant to cross the road until one had passed. After all, early tests with such vehicles sometimes ended in pedestrian fatalities, and some even failed to recognize wheelchair users. I've watched one haplessly wedge itself between a recycling bin and a phantom obstacle while attempting a fifteen-point turn. Still, on average, I've found them to be no worse, perhaps even better, than human drivers—although this may say more about where I live than the quality of autonomous driving. I'm still careful, but they have begun to fade into my peripheral awareness. Adorned by sensors, these robot taxis, alongside computerized human-operated cars, contribute to the sensory array in cities. Through such smart devices, AI senses and regulates the mobility of people, goods, money, and traffic on our streets. Self-driving cars, like delivery drones and smart cameras, are physical proxies for AI in the built environment, and represent the incursion of intelligent digital systems into the very fabric of our embodied lives.

With the ubiquity of sensing and recording devices in its built environments, smart cities can trap us in entrenched movement channels that suit only some of its inhabitants, or they could observe, foster, and cultivate myriad ways of moving through urban environments. Sometimes the smart city is attuned to the diversity of walking bodies. In Singapore, for instance, senior citizens and residents with disabilities can activate the Green Man + function at many of the city's pedestrian crossings. When an eligible user taps their identifying concession card on the reader at the traffic light pole, it extends the crossing time by up to thirteen seconds. A different game of Green Man, no doubt, than the one with which I opened this chapter; this one embraces technocratic order and the flexibility of computation. Even as something is inevitably lost when bodies are translated into mere information, that translation can sometimes serve as a tool of recognition. When designed with care, computational systems can identify and respond to difference.

Cities are not technologies—not exactly. But like technologies, they often assume and are built for typical users. And increasingly, they are impregnated with digital technologies that imbue them with the capacity to sense, sort, and act upon their citizens. All feet, all bodies, move differently; not meeting certain metrics should not be mistaken for a deficiency to be corrected. Likewise, not all people share the same desires for engagement, visibility, or privacy. When designed around the fiction of a one-size-fits-all citizen, technologies fail to serve our rich humanity. The same is true of smart cities. If smart cities are not built to recognize, distinguish, and serve that difference, citizens could once again be sleepwalking through them. As the Rolling Quads reminded us, every day, through walking, riding, or rolling, we make the cities in which we live. Architect Juhani Pallasmaa has said that "a profound design process eventually makes the patron, the architect, and every occasional visitor in the building a slightly better human being."[18] By attending to the rich, idiosyncratic intelligence of our bodies, we might build digital worlds—actual or virtual—that stretch us to be better.

6

BODY

IT HAPPENED IN 1993, IN THE WILD WEST OF THE EARLY
Internet. In a mansion made of words, a cruel clown named
Mr. Bungle used a voodoo doll to assault a Haitian trickster
spirit, forcing his victim to sexually service him. After flee-
ing to his private chambers elsewhere on the grounds, he
unleashed the voodoo doll's foul magic upon other inhabitants: Moondreamer, the radical Kropotkin, and Snugberry
the squirrel. He compelled them to violate themselves and
one another, his evil laughter ricocheting throughout the
mansion. His nefarious spree only stopped when the wise
elder Iggy arrived, wielding a powerful gun that enclosed
Mr. Bungle in a cage of code—finally foiling the voodoo
doll's supernatural reach. It's a scene that sounds as farcical

as it does horrific. But for its characters, it was profoundly real and painful—even if it only took place in words.

The cyber-rape took place just weeks before the public debut of the World Wide Web, on a multi-user dungeon (MUD) called LambdaMOO. At its core, LambdaMOO was a database masquerading as a world, a program designed to mimic the sensation of navigating a physical space through narration and dialogue on a text-based user interface. After dialing in from their home computers, users crossed the threshold into the mansion, ushered in by descriptions like this: "The closet is a dark, cramped space. It appears to be very crowded in here; you keep bumping into what feels like coats, boots, and other people (apparently sleeping). One useful thing that you've discovered in your bumbling about is a metal doorknob set at waist level into what might be a door. Next to it is a spring lever labeled 'QUIET!'" Players could interact with the world—and with one another—entirely in pithy snippets of text: "I take a cookie from the jar," or "Snugberry slaps Kropotkin with a fish." When they reached a new room, the user's screen would unfurl a description of the room, as well as everyone and everything in it. For many, LambdaMOO was more than a game; it was a second life. Moondreamer, Snugberry, and the infamous Mr. Bungle were not just avatars but deeply inhabited selves—inventions and extensions of their users in a fledgling digital world.

In his well-known telling of this story, journalist Julian Dibbell called LambdaMOO "virtual reality" and described how, through community dialogue and connection, Mr. Bungle's victims and "a cast of dozens turned a database into a society."[1] At a time when immersive technologies such as augmented reality (AR) and virtual reality (VR) have become increasingly sophisticated, this story might seem almost quaintly offensive. Words hurt, to be sure, but can they constitute such a fundamentally embodied violation? For writers and novelists, the immersive capacity of language comes as no surprise. Books have long served as portals to other times and places, although as a technology they invite a different kind of engagement than virtual reality. Choose-your-own-adventure books edge a little closer to this dynamic, offering readers a hand in guiding the story. LambdaMOO and its shinier descendants scaffold interactivity into their architectures, drawing users into making themselves part of the world—to put some skin in the game, as it were. That this encounter, stripped to mere words, could wound so deeply speaks to the remarkably visceral power of virtual worlds. It also reveals the intimacy between virtual space and our very real, very fleshly bodies—a complexity that defies William Gibson's characterization of cyberspace in his novel *Neuromancer* as "a consensual hallucination," a "non space of the mind" separate from the "meatspace" of our bodies.

Though a far cry from VR as it tends to be understood today—a high-resolution, three-dimensional rendering of an environment—this early transgression captures one of immersive technology's key conceits: the body as interface. By subsuming our bodies, VR promises to recede into the environment itself and let the mind roam free. This chapter explores how immersive technologies—from early MUDs to the metaverse, from phantasmagoria to brain-computer interfaces—have grappled with a persistent paradox: the desire to transcend the body, and the impossibility of leaving it behind. Where the previous chapter traced how the foot roots us in the world, this chapter turns to technologies that promise escape—into virtual dimensions of our own making. Virtual realities overwrite the world we know. They promise transcendence, but return us always to our bodies, which remain the sensor, the interface, and often, the battleground.

Technology desires disappearance. When a tool is working as intended, you don't think about it—until it breaks. This kind of disappearance doesn't just require a good tool; it demands skill and practice of the human using it. Like a surgeon with a scalpel or a carpenter with a chisel using their intelligent hands, disappearance is a collaboration between well-made tools and disciplined bodies. Digital technologies push this further still: the ideal tool is one that will completely dissolve, making the human body itself the interface.

XR—encompassing virtual reality, augmented reality, and mixed reality—strives to be that tool. As media become environments, technology fades into nature. Neal Stephenson's novel *Snow Crash* dubbed this imaginary realm the "metaverse," a digital land free from the restrictions of gravity, biology, or geography. But even in this imagined escape, our bodies remain the vessels of departure. These immersive technologies recruit the human sensorium only to efface it, relying ever more deeply on flesh to function even as they pretend to disappear.

Before headsets and haptics, this dream of vanishing machines and liberated minds flickered to life in fiction. One of the earliest blueprints for this vision appeared in "Pygmalion's Spectacles," a 1935 short story by American science fiction writer Stanley G. Weinbaum. The spectacles in question, invoking the Greek myth about a sculptor who falls in love with his own creation, are a pair of goggles through which the wearer experiences "a movie that gives one sight and sound . . . taste, smell, even touch . . . you are in the story, you speak to the shadows and they reply, and instead of being on a screen, the story is all about you, and you are in it."[2] Rooted in George Berkeley's idealist philosophy, the story imagines reality as a cascade of sensory impressions— "the windows of your five senses"—through which we never perceive things directly, only the sensations they evoke. By appealing to those senses, the goggles don't just simulate the world, they conjure its felt reality.

Here, immersive technology reveals an ancient ambition to hack the circuit of feedback between our bodies and our environments. As the story unfolds, the protagonist Dan becomes enamored of his beautiful host, Galatea, who asks him, "Does the real world seems strange after that shadow land of yours?" In that question lies the allure of the virtual—that our world is but a shadow play of another, more vivid one awaiting behind the veil: "He had forgotten the paradoxes of illusion; this was no longer illusion to him, but reality itself." In this, "Pygmalion's Spectacles" captures the essence of immersive technology's spell: a section of the senses that transmutes our bodies into instruments of transportation. Unlike the smart city, which embeds technologies in the environment, virtual reality gathers it close to the skin. In the smart city, the world is the interface for our bodies; in VR, the body becomes the interface for the world. Either way, we are building new digital worlds through the machinery of our imaginations.

Text-based worlds like LambdaMOO demonstrate that you don't need graphical immersion for a world you can really live in. But you do need to feel invested in the community, and one time-honored way to do this is to participate in building it. The 2003 launch of Linden Lab's *Second Life* gave a name to these inhabitable digital realms. *Second Life* is a multiplayer virtual world where people can create avatars of themselves. Through these avatars, users navigate through a shared, computer-generated 3D world and can interact with other others. In this respect, *Second Life* is not especially unique. Immersive video games have captivated legions of gamers for decades—from "first-person shooters" like *DOOM* to the notorious *Grand Theft Auto*.

Unlike many 3D worlds that players might wander, however, *Second Life* entrusts the work of world-making to its residents. Although its infrastructure is engineered by Linden Lab, it is a world of its own making, truly a product of collective imagination. From the pious to the perverse, the mundane to the ridiculous, users can build anything they can imagine. The platform is home to massive public art installations, a gaudy mansion that defies gravity, a neoclassical palace adorned with Ukrainian flags, and a dark island filled with little vampire children, to mention a few of its manifold virtual zones.

Second Life makes real the wildest and weirdest of imagined places. For many people, it is more than a game. It's home: a social hub, a canvas for creativity, a refuge from the boredom of family life, liberation from the body hunched over the computer, and an incarnation of the phenomenon sociologist Sherry Turkle calls being "alone together." People have fallen in love in *Second Life*, sometimes while their first love sat just a room away in the real houses where their actual bodies resided. Indeed, *Second Life* can be an escape for people who feel trapped by the realities of their bodies. Some people live fabulous lives as fantasy architects, chimeric rock stars, and digital couturiers, all creatures of a virtual menagerie inhabiting a world where anything is possible.

Second Life even has its own currency, lindens, with which citizens can buy and sell virtual property, from clothes to cars. Lindens can be converted into real money, legal tender that can be exchanged. In 2007, a player called Stroker Serpentine sold an entire virtual city for $50,000. Built on the trade of such virtual goods, its economy is valued at $650 million a year. Though eclipsed in popularity by newer platforms like *Roblox*, *Fortnite*, and *Minecraft*, *Second Life* still sees tens of thousands of users on the platform at any given time.

To its residents, *Second Life* is more than a virtual veneer over our reality. With its own economy, its own society, its own extravagant fashions, its own subcultures, and its own identities, *Second Life*'s abiding existence seems to fulfill the promise of its name and of its immersive forebears. It prototypes one version of Pygmalion's spectacles, offering a reality more tantalizing, attractive, and fantastical than the dreary one many of us experience every day in the bodies we were born with. While *Second Life* has endured as a user-centered, creator-driven virtual world, it has been outstripped in scale by newer platforms, and bolstered by Big Tech's investment in immersive tools like Meta's VR headsets and Snap's AR glasses.

Fueled by increasing computational power, with high-definition graphics and surround sound, contemporary XR technologies are impressively glossy, if sometimes glitchy. But the desire to inhabit alternate realities is not new. Virtual reality finds its lineage in nineteenth-century experiments with immersive spectatorship, all of which sought to position our bodies in closer proximity to imagined worlds. As these experiments reveal, we have been eager to cede our bodies to immersive worlds so we can escape this one. This wish to escape our own bodies is ancient, predating even the earliest of these immersive experiments. As writer Meghan O'Gieblyn argues in *God, Human, Animal, Machine*, this longing has deep theological roots. Judeo-Christian metaphors of the soul and afterlife seeded modern thinking about the mind-body split. Today, these same metaphors echo in the trans-humanist hope of mind uploading and digital immortality. Whether through goggles or code, this yearning continues to animate experiments in virtual embodiment.

Like death photography and voice recordings of the dead, immersive media have long flirted with the supernatural. Sixteenth-century accounts describe how witches and magicians used proto-cinematic technologies such as camera obscuras and magic lanterns to conjure spirits and demons in necromantic rites. Johann Georg Schröpfer, a notorious German charlatan, was infamous for using hidden magic lanterns to raise ghosts at seances where attendees were allegedly fed hallucinogenic punch. Many of the earliest projections were of demons and hell, apparitions from worlds humanity had long feared but never seen. Emanating with supernatural meaning, these projection technologies gave depth and reality to these imagined worlds. The earliest known reference to the magic lantern is a page of sketches circa 1659 by Dutch scientist Christiaan Huygens, considered one of the technology's possible inventors, depicting a skeleton removing its own skull. In these performances, showmen used strategies such as rear projection to obscure the technology, constructing experiences of immersive immediacy unencumbered by reality.

By the late eighteenth century, these techniques had matured into the phantasmagoria, a genre of horror theater that combined projection, spooky special effects, surround sound, and other sensory stimuli to create enveloping environments. Aided by multiple magic lanterns, skilled operators cast shimmering demons, dancing skeletons, and floating ghouls onto gauzy screens hanging in darkness. Phantasmagorias were popular in much of Europe well into the nineteenth century, as showmen perfected their spooky craft and innovated using the technology to create ever more immersive experiences. One of the most well-known pioneers of phantasmagoria, Étienne-Gaspard Robert (stage name: Robertson), was said to take pride in terrifying his audiences; the more they cowered in fear from the apparitions in front of them, the better. Many modern haunted houses, including Disneyland's Haunted Mansion and Phantom Manor, derive their effects from this phantasmagoric legacy. Then as now, the spectacle depends on its sleight of hand: the more invisible the machinery, the more immersive the magic.

Around the same time, a more majestic immersive medium rose to prominence: the painted panorama. Whereas phantasmagorias hurled audiences into hellish realms, the painted panorama invited aspiring travelers to behold the grandeur of the greater world. Designed to flood the senses, these colossal paintings encircled viewers in 360-degree vistas of battlefields, cityscapes, ancient ruins, or global wonders. Installed in purpose-built rotundas, panoramas resurrected history at glorious, terrible scale. In Europe, the Napoleonic Wars were a panoramic staple; in Asia, scenes like Tokyo's "Attack on Pyongyang" and China's "Lugouqiao Incident" evoked national memory. The panorama collapsed time and space into a single, enveloping view. These early immersion machines prefigured today's digital realities—worlds remade in paint, light, and the galaxy of the human sensorium.

Innovations in panorama design enhanced sensory stimulation and layered visual details to heighten the illusion of immersion, presenting convincing worlds to audiences to observe through the windows of the senses. One of the most ambitious of these was The Trans-Siberian Railway Panorama, which won a gold prize at the 1900 Paris Exposition. Commissioned by Compagnie Internationale des Wagons-Lits and shown in the Siberian section of the Russian Pavilion, the panorama simulated stages of the journey from Moscow to Beijing on the Trans-Siberian Railway. Contributing to the verisimilitude of the experience, visitors would sit in one of three seventy-foot-long luxury railway cars, which were replete with bedrooms, dining rooms, and even saloons. To create the illusion of movement, four layers of scenery reeled past the stationary spectators at different speeds and distances. Closest to the car, a fast-moving belt with rocks and boulders turned in the foreground. Slightly farther away, a low canvas painted with brush and shrubs moved more slowly. Behind that, two larger canvases, one with distant scenery and the other with mountains, cities, and woods lay in the background. The effect was a dynamic landscape, creating motion parallax that mimicked depth and movement. Gazing out of the window, passengers felt themselves transported. These early immersive formats were not merely feats of illusion—they were portals. With such technology, one's body didn't actually need to go anywhere to be somewhere else. In this respect, immersive technologies have always indulged a fantasy to be anywhere but here—a desire that proved prescient in today's virtual voyages.

As a form of transportation, railways would provide a visual shorthand for the kind of journey these immersive efforts offered their spectators: mobility without displacement. This symbolic power resonated across other media of the era, particularly stereography, where images of railways and distant landscapes became staples of optical excursions. Invented almost concurrently with photography, the stereoscope was based on a curious illusion described by British scientist Charles Wheatstone in an 1838 paper. He discovered that if someone looked at two images depicted from slightly different perspectives, then viewed each through a different eye, the brain would assemble them into a three-dimensional view. Based on Wheatstone's tabletop prototype, the handheld stereoscope made its debut in the marketplace in the mid-nineteenth century. With its capacity to render real-world scenes in vivid three-dimensionality, the stereoscope became a portal to many places. The second half of the nineteenth century saw a craze for these devices, so portable yet so powerful in their capacity to teleport people to distant destinations.

Being able to see images in three dimensions was a major

innovation in visual technology; after photography, it was the most seismic. According to American physician and writer Oliver Wendell Holmes Sr. in an 1859 *Atlantic* essay, "The first effect of looking at a good photograph through the stereoscope is a surprise such as no painting ever produced. The mind feels its way into the very depths of the picture. The scraggy branches of a tree in the foreground run out at us as if they would scratch our eyes out." He waxes lyrical as he gushes on in remarkably tactile language: "Oh, infinite volumes of poems that I treasure in this small library of glass and pasteboard! I creep over the vast features of Rameses, on the face of his rock hewn Nubian temple; I scale the huge mountain-crystal that calls itself the Pyramid of Cheops."[3] Like the goggles in "Pygmalion's Spectacles," the stereoscope constructs a dimensional reality that can almost be touched.

In Europe, stereographic imagery was often of castles, cathedrals, and other ancient landmarks. In the United States, stereographers captured the country's extraordinary and stunning landscapes, from the arid high desert to lush waterways. Scenes from abroad were also particularly popular, spawning legions of stay-at-home tourists and armchair anthropologists. Underwood and Underwood, then the world's largest producer of stereographs, were at one point producing 10 million images every year. They released themed box sets, with their travel series especially in vogue. The stereoscope achieved broad popularity in Victorian culture, shared by young and old, rich and poor alike. Disappointed by what he saw as its crass pandering to the senses, Charles Baudelaire lamented "a thousand hungry eyes . . . bending over the peepholes of the stereoscope, as though they were attic-windows of the infinite."[4] For some, the stereoscope was a democratic marvel; for others, it was a vulgar machine for turning wonder into a commodity.

I had one when I was little. It was a red View-Master with an orange tab I could press with my thumb to flip through different reels. Each reel was a thin cardboard disc with seven stereoscopic pictures of places I'd never been and people I'd never met. The View-Master was introduced at the 1939 New York World's Fair, a century after the debut of the stereoscope. Intended to replace scenic postcards, these images— and the act of looking at them—seem to embed spectators within the representation. From the 1940s through the 2000s, View-Master produced over a billion reels. These reels encompassed global peoples from around the world, life on every continent, cartoons, popular culture, plant taxonomies, art, and science. View-Masters were used for work and play, by children and scientists, educators and artists.

Despite their varied mechanisms, these immersive ancestors all turn on a longing for unmediated experience viscerally felt. It is this very yearning that makes us vulnerable to the predations of corporate entities seeking to immerse us in their proprietary operating systems. Just as smart cities induct human bodies into environmental systems optimized for efficiency and prosperity, immersive technologies of the twenty-first century install users into corporate platforms that dictate how we live, work, and experience the world.

New headset innovations perform the same kind of visual sleight of hand, retooling stereoscopic principles for 360 video—in service of new modes of labor, consumption, and socialization defined by the platforms that house them. The Apple Vision Pro, for example, claims to inaugurate a new era of spatial computing by "seamlessly blending" digital content with the physical spaces we're in. In so doing, they aim to integrate not just the digital world, but the logics of Apple's operating system, with the meatspace we inhabit every day. It is the next phase in the assimilation of our bodies into operating systems, extending iOS's quiet colonization of our pockets, palms, and psyches—of anyone who uses an iPhone or iPad.

In 2014, Facebook (now Meta) acquired Oculus, a VR headset company. Three years later, they unveiled an ambitious project: a noninvasive commercial brain-computer interface (BCI) that would allow users to type with just their thoughts. They recruited scientists from several storied scientific institutions and universities in service to the project. One multiyear collaboration between Facebook Reality Labs and University of California San Francisco's Chang Lab, Project Steno, was a research effort aimed at developing a system for translating brain activity into words. The lab's research entailed implanting BCIs—systems with implanted electrodes that can detect brain activity to train a computer system—into the brains of people with speech impairments or paralysis, in this case, a man who lost his ability to speak after a stroke sixteen years prior. Over a year, that man spent twenty-two hours training a system to recognize specific brain patterns while attempting to speak words and sentences, producing a language model that could decode words from brain activity with some success, albeit very slowly. Reality Labs also prototyped an external headband for measuring brain signals, kind of like a pulse oximeter for the brain. They've since placed this particular model on hold, pivoting to a device with a clearer pathway to market. In 2019, they acquired CTRL-Lab, a start-up developing a neural interface wristband that uses electromyography (EMG) to detect motor nerve signals, translating intention into action through subtle muscle movements.

Among a slate of other "natural" and "intuitive" interfaces Reality Labs is working on, these tools aim to short-circuit the cycle of feedback between our bodies and our digital worlds. Their work is animated by the drive to erase all signs of mediation—to make technology recede into either our bodies or our environments. A brain interface—controllable by thoughts and able to make a user feel they are in a different reality—promises to manifest the vision of reality that Pygmalion's spectacles offered, displacing this one. Calling this "the magic of presence," Reality Labs hypes connection as the primary motivator of their R & D. But the connection they shill isn't truly to one another; it's a constant tether to their digital infrastructure, where multitasking masquerades as presence. At their core, brain-computer interfaces might seem to be the inexorable and logical conclusion to the history of immersive media. After all, technology longs to disappear. So realized, what actually fades from view are the troubling information architectures that induct users into patterns of consumption and social activity structured by the large companies that code and build them. As interfaces become more invisible, they exact their pound of flesh from our living bodies. Reality Labs envisioned use cases for electromyographic wristbands based on gestures that could be executed "regardless of where you are or what you're doing, while walking, talking, or sitting with your hands at your sides, in front of you, or in your pockets."[5] Performed in public but perceptible only to machines, these silent gestures sever interaction from the shared spaces of real life, splitting attention and privatizing presence.

Such initiatives build infrastructure for an imagined future in which BCIs and VR headsets become the new frontier of socializing. Must we don, ironically, ever more hardware in order to retreat into another imagined world? While many people have tried some version of XR (Pokémon GO counts!), headsets are few and far between, too expensive and inaccessible to most. VR and BCIs can seem inconceivably cutting edge, a distant future for many whose technological interactions are far more mundane.

To be honest, while waiting for this future to arrive, I hadn't expected it to be so boring. The Apple Vision Pro, touted upon its debut as being the next big thing in spatial computing, was advertised with users surrounded by life-size spreadsheets, web browsers, and other productivity tools. Of course, its other immersive features—hi-res photos and films—are lauded as magical; you can watch a movie at IMAX scale while confined to an airplane seat. But frankly, the prospect of wrangling an Excel budget as big as I am sounds like a nightmare.

This embrace of extended reality as a space not to escape, but rather to enlarge, our desktops, is perhaps the miserable conclusion of the world we've built. In this respect, these glasses make literal our metaphorical swallowing by the productivity tools and other digital platforms we use. As philosopher and cognitive scientist David Chalmers puts it, "VR devices aren't illusion machines; they're reality machines."[6] For most people, virtual reality might conjure images of futuristic worlds that are yet to arrive. But the virtual reality most of us inhabit are the social networks, telecommunications tools, and everything that is available on the personal devices in our pockets. These digital experiences are being overlaid on our lives in prosaic and subtle ways. For many of us, our daily realities are quietly, constantly augmented: we look at dynamic maps to guide us rather than the terrain underfoot; we document our lives through the lens of apps that perfect our blemishes; we conduct much of our professional lives through videoconference. Like augmented reality, these ubiquitous apps, platforms, and functions are an ever-present digital veneer over the faces of our existence. These second lives are far harder to disentangle from than the patently fictional VR experiences that have captivated their users.

We're already living in mixed reality. Our bodies are entangled in a dance with data: computers track our keystrokes, footsteps, and heartbeats; they reproduce and organize our movements; intelligent systems choreograph our journeys large and small; we socialize through electronic sound and through avatars in virtual spaces. Extended reality technologies don't simply show us other worlds; they clarify the one we're already in and reveal how deeply our lives are intertwined with computation. At the same time, we cling to old myths to make sense of new machines. Why are we so invested in these tales of transcendence and escape? These desires at the heart of the history of immersive media have much to tell us about the present and future of virtualized digital bodies.

I'm on Zoom several times a day for work. The beginning of most calls is so familiar, as if it is scripted, as if digital ghosts call out to each other across an invisible veil: "Hello, are you there, can you hear me? I can't hear you. I can't see you." Like the seances and phantasmagoria that prefigured immersive media, these regular Zoom mishaps rehearse a horror at the heart of the medium. The ghost in the machine is us, after all. We are the pale reflections we see in the virtual mirror onscreen. Defined by Big Tech, our avatars are now far less interesting than the weird and wonderful characters on LambdaMOO and *Second Life*.

We burnish our digital images (I'll admit that mine is lightly airbrushed by the Touch Up My Appearance option in my Zoom preferences). We feed ourselves to the technologies we use, seeking to transcend the limits of our bodies and minds. We are spit out as ghosts of the platforms that puppet us. The term "ghost in the machine" has been used as a crude and derogatory jab at Descartes's mind-body dualism—the idea that our minds animate our bodies like spirits inhabiting a shell. One version of the body digital inverts this: the mind floats free, divorced from our bodies and assimilated by platforms. But we are not disembodied minds. We are deeply rooted in flesh, blood, and bone. Any future worth building must remember that.

In 2021, Facebook rebranded itself as Meta. Invoking the metaverse, this name signaled the company's ambitions to create an immersive platform in which all of life's activities—social and professional transactions alike—might be conducted. In 2024, Meta debuted Ray-Ban smart glasses: AI glasses that allow wearers to capture photos and videos, call people, "translate a sign, summarize a food menu or remember where you parked your car."[7] Rendering the technology itself invisible, the product aims to further naturalize the merger of everyday life and extended reality. They're just cool sunglasses, you see.

That year, Meta's annual report revealed that it had invested $19.9 billion in Reality Labs, contributing to a cumulative investment of over $80 billion in AR and VR—and they're looking to ramp up these investments further. The goal appears to be nothing less than building new infrastructure for everyday life, essentially becoming reality's operating system. From electromyographic wristbands to Meta's prototype Orion AR glasses, wearable R&D pursues a cloak of invisibility—technology that vanishes, even as its grip tightens. Meta Reality Labs peddles this vanishing act as a return to the human: building "natural" and "intuitive" interfaces that supposedly center the user. But the truth is that the more these interfaces disappear from our awareness, the less we are awakened to how they choreograph and constrain our experiences. What would genuinely centering the human be like?

The cyber-rape in LambdaMOO, and its aftermath, unfolded in the Wild West of the Internet, a more fluid era when communities took shape through experimentation and improvisation. Want a personal website? Code it yourself, line by line, in HTML. My own teenage website was cobbled together from borrowed code and basic markup—rudimentary, to be sure, but very much mine. Now, several social platforms make quick work of virtual identity through polished templates and drop-down menus. A handful of social platforms streamline self-expression, ossifying the way people connect to one another in predetermined ways. Far from this Wild West, many of us now instead reside in an Internet defined by walled gardens.

A walled garden has historically referred to a literal garden—enclosed by high walls that shelter delicate plants, create microclimates, and shield the space from prying eyes. Such a garden is easier to design and cultivate, for its architects tame the land by controlling the environment itself. In technological parlance, the term now describes a closed software system where every element, from content to inter-actions, is controlled by the platform's architects.

Closed platforms—like Instagram, Facebook, or Twitter—are walled gardens in that they are ecosystems that control all variables, determining how users interact, what they see, and how they're seen. Participation demands sub-mission to community structures that govern those interac-tions—however shiny, however pleasurable. Consider the Facebook "like," the Instagram "heart," and the "friends" list: whatever the algorithm offers as options become the ways users engage with one another. As writer Zadie Smith has argued, Facebook reduces people and their endless quirks and idiosyncrasies to a set of data.[8] These profiles are not your friends.

As immersive technologies career headlong into the vor-tex of innovation, these concerns remain urgent. The luster of AI-powered eyewear—goggles that survey our environments, anticipate our needs, dim the noise—has been polished by BCI prototypes. Neuroadaptive systems that bridge BCI feed-back with augmented reality detect cognitive or emotional strain and subtly recalibrate the visual field: fading distrac-tions at stress peaks, sharpening outlines when focus dips. While such responsive integrations might feel miraculous, they turn our neural signatures into data, harvested in real time. The walls grow ever higher as the boundary between self and system melts into the wetware of our own minds.

Rather than fostering connection between people by dissolving the technology into our bodies, sleek AI glasses or BCI force us further into these walled gardens, raising their walls so we forget that there is a world beyond them. Building communities and being in community takes more than immersion. As the LambdaMOO story reminds us, a reality worth inhabiting—augmented, virtual, or tangible—requires a politics of interaction, a commitment to one another, and a recognition that our messy bodies mean something, however much they are projected into pixels.

7

MIND

BEFORE SIRI, ALEXA, OR CHATGPT, THERE WAS ELIZA.

Men are all alike.

IN WHAT WAY.

They're always bugging us about something or
other.

CAN YOU THINK OF A SPECIFIC EXAMPLE.

Well, my boyfriend made me come here.

YOUR BOYFRIEND MADE YOU COME
HERE.

He says I'm depressed much of the time.

I AM SORRY TO HEAR YOU ARE
DEPRESSED.[1]

If you heard this conversation, you'd be forgiven for thinking you might be eavesdropping on a psychotherapy session. It is, however, an exchange with the world's first chatbot (then called a chatterbot), Eliza. In 1966, an MIT professor named Joseph Weizenbaum created Eliza at the university's Artificial Intelligence Laboratory. Published by its creator in a journal article explaining how the chatbot worked, this snippet captures a typical encounter with the conversational agent. A user would talk to Eliza by typing a message on an electric typewriter that was connected to a mainframe computer. Momentarily, it would reply.

Named after Eliza Doolittle, the cockney flower girl in George Bernard Shaw's *Pygmalion* who learns how to speak like—and thus pass for—an aristocrat, Eliza was an early natural language processing program designed to simulate a human conversational partner. The software applies a set of pattern-matching rules to search for keywords and generate responses that, as you can see from the exchange above, suggests comprehension. One of the most famous variations of these scripts was called DOCTOR, which was engineered to mimic a psychologist of the Rogerian school, a form of patient-centered therapy. According to Weizenbaum, the format of the psychiatric interview allowed the bot to "assume the pose of knowing almost nothing about the real world."[2]

Eliza was remarkably convincing in how it seemed to

understand its interlocutors. Weizenbaum observed that "some subjects have been very hard to convince that Eliza (with its present script) is not human."[3] In a later article, he told the story of his secretary asking for some time with Eliza, shortly thereafter asking Weizenbaum to leave the room for privacy. He emphasized the chatbot's capacity to perform comprehension, saying, "I believe this anecdote testifies to the success with which the program maintains the illusion of understanding."[4] As one of the first computer programs that convincingly simulated a human presence, Eliza aroused what came to be known as the "Eliza effect," which refers to our tendency to project human characteristics onto artificial systems such as computers, machines, and AI. In its day, Eliza was a sensation. *The Boston Globe* even sent a journalist to its typewriter, publishing an excerpt of the interaction. The chatbot roiled the world, becoming one of the most iconic programs in computing history. Eliza would even go on to converse with another chatbot, Parry, coded to simulate a person with paranoid schizophrenia—a prophetic warning of the manifold machine-to-machine exchanges that drive much of our social, cultural, and technological infrastructure today.

In the more than half century since Eliza's birth, large language models—contemporary incarnations of chatbots— promise to surpass this algorithmic progenitor. Large language models, or LLMs, are bleeding-edge innovations in generating and understanding human language that draw on enormous information corpora far beyond the scale of the individual human mind. ChatGPT, Gemini, Claude, and Grok are some of the best-known LLMs, which feast on massive databases and colossal amounts of energy to cheerfully complete tasks, from research to image generation, for their ever-increasing user bases. And yet, in 2023, Eliza beat OpenAI's GPT 3.5 in a Turing test study, a reminder that believability is not just a function of computational power, but of human projection and context. As the field of artificial intelligence has matured, this grande dame of software continues to be a revealing yardstick for how we understand both machine thinking and our own. Early chatbot exchanges like Eliza's were more than technical novelties; they were self-reflexive dialogues about the health and nature of the human mind.

Like nearly all major technological advancements, LLMs are at once lauded and vilified. Current cultural conversations about these tools rehearse familiar tensions: between promise and peril, productivity and obsolescence, automation and authenticity. Teachers bemoan the canned robot-authored essays they will receive; artists and writers mourn the impending demise of their careers; acolytes trumpet unprecedented boosts in productivity. Large language models are evolving at breakneck speed and are rapidly being deployed in every sector that computers touch—that is to say, almost all of them.

At the time of writing, OpenAI had just struck a deal with the entire California State University system to bring ChatGPT to over five hundred thousand students and faculty across twenty-three campuses. The initiative aims to embed ChatGPT into every facet of university education, from administrative tasks to tutoring. Similar integrations are unfolding globally: in China, major universities, including Zhejiang University, Jiao Tong University, and Renmin University are similarly incorporating homegrown AI platform DeepSeek into teaching, research, and campus operations. Critics lament the risk this widespread adoption poses to literacy and cognitive development. After all, for students, completing research, writing, thinking, and problem-solving tasks is far less about demonstrating mastery than acquiring knowledge and skills through practice and effort. Use it or lose it (or never gain it)—it's hard to learn when skipping the hard part. Classrooms provide staged practice and the space to make mistakes. Educators I know have struggled with teaching in this uncertain new era, trying to conceive of ways to assign AI tools in classrooms in ways that foster learning rather than circumventing it entirely.

However, in many ways, these AI innovations continue a project initiated by the technology of writing itself. From the invention of writing to the birth of computing, humans have long made tools that think with and for us. These technologies are not just cognitive aids; they shape what we know and how we come to know it. To grasp the stakes of AI today, I return to writing not merely as a record of thought, but as one of its oldest architectures. Writing gave ideas external and enduring form beyond speech and mind; in so doing, it both extended mental processes for individuals and granted community access to an always-growing library of concepts. This tension between embodied extension and detachment marks all our relationships with technology—and provides an avenue for imagining more human ways of being with and through our tools. In the first chapter of this book, I talked about writing as a physical practice: a choreography of hand and mind, rehearsed and mundane. This chapter comes full circle by returning to writing—liberated not only from the human hand, but from the human mind.

Across the Internet, people use LLMs to write emails, generate articles, pump out code, create poems, plays, and other literary forms. Gorged on literary data, including scores of articles, and reams of books from Shakespeare to Dr. Seuss, LLMs can make for remarkably fluent writers. Its creative output, however, remains blandly obvious facsimiles.

In 2024, writer Curtis Sittenfeld participated in an experiment orchestrated by *The New York Times*, essentially a literary Turing test between herself and ChatGPT. In this contest to see who could write "a better beach read,"

Sittenfeld and *The New York Times* invited readers to vote on its creative parameters. Armed with a set of themes, textures, and features—lust, regret, kissing, middle age, flip-flops—Sittenfeld and ChatGPT went to work, the latter through a set of prompts asking it to write a one-thousand-word story "in the style of Curtis Sittenfeld." Sittenfeld is one of myriad fiction writers whose novels were used to train ChatGPT, with neither permission nor compensation, so the LLM was familiar with her style, to say the least. The two stories were offered to readers of *The New York Times* to judge for themselves, with a postscript from Sittenfeld denoting which story was hers, and listing some idiosyncratically human things she'd done as part of her process, such as driving to the park where she set her story, soliciting feedback from family members, and dawdling (ChatGPT's process? Spit it out in seventeen seconds). To Sittenfeld, the distinction was clear: ChatGPT's story was boring, clichéd, and shallow. Through my very scientific assessment of the article's comment section, I concluded that this was obvious to anyone who liked to read.

Other writers have integrated AI into their processes with less skepticism, positioning the technology as a form of intellectual and artistic augmentation. In 2019, science fiction author Chen Qiufan published a short story, "State of Trance," that contained passages generated by an algorithm trained on his own writing. This story would go on to win a Chinese literary contest presided over by an AI judge, beating out Nobel Laureate Mo Yan by a slender margin (scoring 0.00001 of a point ahead, if literature can be so quantified). Fittingly, "State of Trance" is about a man's journey to return a library book to the Shanghai Library as the world collapses around him, humanity as we know it is approaching its end, and artificial intelligence is taking over—the dying gasps of writing and thinking as we know it. Rather than a literary ouroboros, this story gestures at the need for reimagining the future of writing in a world inexorably changing in the wake of AI.

While a long road remains until AI might pose an existential threat to creative writers, it has already begun to serve as a writing tool. And like all writing tools, it challenges the notion that the skull marks the border of the human mind. Writing systems of all kinds have reified the intangible; collecting ideas into archives, they are the building blocks of external memories that can turn thinking into an interchange between mind and database. As prosthetic memories, the earliest forms of writing can be understood as ancestors of the information retrieval and analysis that define contemporary machine learning. AI offers new ways of working with databases—new ways of thinking and creating. By extending humans' cognitive capacities, writing helped to sustain profound cultural transformations. AI may yet do the same.

But as the uneven legacies of literacy suggest, the stories we tell with our writing tools are just as consequential as those tools themselves.

In 2017, the artist and self-styled "gonzo data scientist" Ross Goodwin drove a Cadillac from New York to New Orleans, its trunk packed with AI software trained on three literary corpora, including science fiction, poetry, and "bleak" literature, as well as Foursquare location data. Goodwin had coded the algorithms himself and curated an archive of hundreds of books to ground the AI's linguistic matrix and aesthetic sensibility. And with Google's support, he rigged the car with a microphone, clock, GPS unit, and rooftop camera.

As Goodwin piloted the car south, the system of neural networks synthesized sensor data in real time, generating text in a poetic idiom evocative of Jack Kerouac's spare stream-of-consciousness prose. The result was sometimes lyrical, sometimes surreal. The novel begins, "It was nine seventeen in the morning, and the house was heavy." An algorithmic recognition of a Foursquare location produced the sentence: "Eagles Nest Diner: a American restaurant in Goldsborough or the Marine Station, a place of fish seemed to be a man who has been assembled for three days." Each line was printed on long rolls of receipt paper, alongside time stamps of when they were generated, and compiled into *1 the Road*, which Goodwin claims is "the longest novel in the English language."[5]

1 the Road, at once a literary experiment and a performance, highlights the complexities around assigning authorship when artificial intelligence is involved in the creative process. The algorithms wrote the text, but they were shaped by Goodwin's choices: his code, his data corpus, his route. With a car for a pen, Goodwin and his AI staged a twenty-first-century buddy road trip, chronicling the journey of a writer on the move and in the making—not just behind the wheel but entangled with machines, environments, and data.

From the start, computers have been understood in relation to the human mind. Apple co-founder Steve Jobs, fabled for his unrelenting vision and pursuit of personal computing, characterized computers as a tool for amplifying human efficiency, "the equivalent of a bicycle for our minds."[6] The chatbot Eliza, with its mirroring dialogue, invited users to encounter their own thoughts refracted through code. From its beginnings, AI was conceived within the terms of mental faculties and envisioned as a form of intelligence that does not simply ape but also augments the human mind. In 1963, the same year MIT invited Weizenbaum to join the faculty, the university launched Project MAC—an acronym that meant, among other things, "machine-aided cognition"—with a

$2.2 million Pentagon grant. Its premise was simple, radical, and foundational: thinking need not be confined to the brain.

In their landmark 1998 paper "The Extended Mind," philosophers of mind Andy Clark and David Chalmers asked, "Where does the mind stop and the rest of the world begin?" Their answer has become one of the most influential articulations of the extended mind thesis, which rejects the conventional view that the mind resides solely within the brain, stopping at skull and skin. Rather, they proposed that cognition arises from the dynamic interplay of brain, body, and tool. A pencil, a notebook, or a computer screen can become so integrated into our mental processes that they functionally bring about our cognitive abilities as much as our brains. The mind, in this view, is porous: it reaches into the world, and the world reaches back. Cognition, then, is not contained but distributed—emerging from an ecology of brain, body, and environment.

By externalizing information outside of the human mind, writing systems—from handwriting to generative AI—install new elements into this dynamic ecology. Granthika, a storytelling software start-up spearheaded by the novelist Vikram Chandra, puts this understanding of extended cognition into practice. Granthika is an intelligent system and integrated writing environment imagined as a writer's assistant and bookkeeper—an external brain. Among other tools in development, the software helps fiction writers build and track complex worlds and timelines. It takes care of the grunt work so writers can focus on narrative elements like themes, plot, and character.

According to Chandra, Granthika doesn't just store text; it builds knowledge. Each entry generates a semantic relationship that reflexively updates the story's internal logic. Over time, Granthika assembles an ontology of the novel's universe. Using classical first-order reasoning, it constructs a behind-the-scenes model of time, forging meaningful connections between people, places, events, and ideas. A partner in imaginative continuity, the software aims to free writers for the ephemeral work of craft.

By delegating the work of world-building to a computational intelligence, Granthika serves as a cognitive extender for creative writers. It incarnates a symbiotic relationship between the creative writer and the intelligent database. Similarly, *1 the Road* distributed the labor of writing across human and machine actors, although Goodwin flipped the script and positioned his AI as the thinking and sensing center of the process. In both projects, authorship is neither singular nor static. Against traditional notions of the lone author, they conceive of creative writing as the product of humans coupled with technologies.

But rather than a radical rupture with writing history, these experiments enact the next logical step in the evolution of writing as a mnemonic technology. For all the advanced information systems humans have developed, writing remains our principal technology for gathering, storing, retrieving, manipulating, communicating, and sharing information. AI systems reconfigure that tradition.

In an era when keyboards and touchscreens mediate much of our communication, earlier forms like handwriting emanate an almost natural aura. As organic as handwriting may seem, it is a technology that also wrought profound transformations on communication, cognition, and culture. Before the invention of writing, the spoken word lived in the evanescent spaces between people, evaporating as soon as they were spoken. Writing captures fugitive speech in flight and reifies it in a visible and enduring form. Whereas a conversation had to take place between two living speakers, written words can exist beyond the presence and lifetime of the scribe.

On the astonishing nature of books, astronomer Carl Sagan once marveled, "One glance at it and you're inside the mind of another person, maybe somebody dead for thousands of years across the millennia, an author is speaking clearly and silently inside your head, directly to you. Books permit us to voyage through time to tap the wisdom of our ancestors."[7] For these reasons, Plato famously condemned writing in *Phaedrus*, saying it would erode the human capacity to remember and breed forgetfulness.

From grocery lists to encyclopedias, writing extends the human mind by offloading the burdens of memory, storing and retrieving information outside the body. Writing is a technology that allows us to outsource individual and collective memory. By sustaining the creation of informational archives that can be referenced, literacy made possible new forms of interaction with language. New techniques of information storage afforded the structured accumulation of knowledge. Once formulated, information can be reformulated with greater and greater precision. In this way, literacy laid the groundwork for the disciplines of logic, philosophy, and science in general—the knowledge infrastructures that would, centuries later, give rise to AI.

Beyond these general observations, scholars of literacy and orality—that is, cultures with no knowledge whatsoever of writing—have long debated how writing transforms modes of thought as well as modes of communication. In the 1960s and 1970s, thinkers like Walter Ong argued that oral cultures rely on rhythmic, embodied strategies to carry memory across generations. In his influential book *Orality and Literacy*, Ong related the syntax of language to the structure of thought, treating language as a kind of memory architecture. His claims, drawn from canonical oral texts such as the Bible and *The Odyssey*, helped frame literacy as a transformative force for human consciousness. This view has since been challenged for its oversimplification of "primitive" oral cultures and literacy's totalizing effects. Because of the complexity and plurality of writing systems, scholars today describe a universe of multiple literacies, attuned to the particular cultures from whence they emerge.

These questions about how different literacies reshape thought and memory are given flesh and feeling in Ted Chiang's speculative short story, "The Truth of Fact, the Truth of Feeling." Told in two interwoven tales—one set in the near future, the other in the colonial past—it explores how new literacies don't just change how we communicate, but how we know, remember, and belong.

In the future narrative, an older journalist investigates a new technology called Remem, a lifelogging tool that allows users to index, search, and replay video recordings of their daily lives. Despite his skepticism, he begins to use Remem to revisit his own memories. As he sifts through archived footage, piecing together memories from his own and others' lifelogs, he confronts long-held illusions about his role as a father. What begins as a fact-finding experiment becomes a reckoning with emotional truth, one that forces him to revise the story he's told himself about who he is and who he's been to his daughter.

The second narrative follows Jijingi, a young man from the Tiv oral culture in Nigeria, as he learns to read and write under the tutelage of a European missionary. As Jijingi studies colonial and religious texts, literacy begins to rewire not only how he speaks, but how he thinks. Jijingi struggles to reconcile his oral culture with the written word. When he turns to written European records to resolve a tribal dispute—pitting oral testimony against written archives—he returns to his clan with evidence refuting their claims. But his elder shrugs and asks, "Have you studied paper so much that you've forgotten what it is to be Tiv?" Jijingi soon comes to understand that while writing may encode facts, it can also estrange its readers from their communities and values.

Throughout the story, readers witness both the journalist and Jijingi's mounting curiosity, confusion, and anguish as these new literacies begin to destabilize the certitude of their own beliefs. Together, these parallel stories examine how tools of memory—be they writing, databases, or digital replay—transform not only how we narrate our lives, but also what we perceive to be true. "The truth of fact" enshrines precision and permanence; "the truth of feeling" honors ambiguity, context, and emotion. These modes of truth aren't in simple opposition, however; they are braided through each person's struggle to locate meaning across different technologies of recall.

"The Truth of Fact, the Truth of Feeling" explores how technologies contour the stories we tell—to ourselves and to one another—and, just as crucially, how the stories we tell shape the technologies we use to tell them. In a final twist, the journalist reveals that he had constructed Jijingi's narrative from fragments of historical fact, imbuing it with imagined feeling, to underscore the limits of artificial memory. "People are made of stories," he reflects. "Our memories are not the impartial accumulation of every second we've lived; they're the narrative that we assembled out of selected moments." Memory, in this light, is not just an archive, but a composition: something we revise and relive through the act of storytelling.

Chiang's story frames both characters as what the journalist calls "cognitive cyborgs," their minds extended by mnemonic technologies old and new. New technologies, and the new literacies they beget, leave both losses and gains in the wake of transformation. Writing, like Remem, is a prosthesis of thought. But while it enhances recall, it also transforms how we make meaning from what we remember. In the Tiv tradition, stories are told with the "whole body, and you understood it the same way," while Remem untethers memory from body. Indeed, these tools may sharpen our access to the past, but they cannot replace the embodied, communal practices through which memory becomes something we live, not just store.

By recording information outside of the human mind, writing systems spawned new ways of sorting and retrieving knowledge. AI and the advent of big data have amplified and automated that reading capacity far beyond the scale of human perception. While Remem remains a speculative technology, its promise of total recall echoes the indexing and pattern recognition at the heart of machine learning. As external and increasingly intelligent forms of memory, AI can deepen the effects that writing systems had in earlier times on the physical borders of the human mind.

Yet, as "The Truth of Fact, the Truth of Feeling" affirms, these technologies—as part of a cognitive ecology distributed across brain, body, and environment—are also embedded within social and cultural worlds. New tools breed new literacies, which can engender nascent forms of knowing, remembering, and telling. Early writing systems scaffolded the emergence of new modes of creativity and communication. So, too, might AI. The stories—and truths—we tell with and about AI can help to tune and integrate these technologies in service of human expression. AI might one day be a partner in writing stories yet unimagined.

From Eliza to Curtis Sittenfeld, chatbot narratives have long been cast as contests, echoing the Turing Test's cold arena: who can pass, who can trick, who can win. These triumphs and failures become a measure of their value and success—and by extension, our own. But perhaps it's time to step out of the ring. When we pit ourselves against self-copies distilled into algorithmic sameness, we don't discover our brilliance, only our banality.

In her prizewinning novel, *Sympathy Tower Tokyo*, Rie Qudan offers another way. The story's protagonist, an architect tasked with designing a prison tower in the center of Tokyo and riven with doubt about the project, turns to an AI chatbot for solace. In interviews, Qudan herself admits to using ChatGPT for emotional and intellectual reflection, saying it prompts her to be more honest than if she were speaking to a person. Short sections of the novel—Qudan says up to 5 percent of it—were written by ChatGPT. While borrowed from ChatGPT, these sections aren't meant to be human, and instead are the responses of the fictional chatbot in the story. In this way, Qudan neither glorifies nor erases the difference. Instead, she stages a relationship, amplifying the human at the same time as identifying a place for machine writing. Human and machine are distinct, but in dialogue.

When it comes to AI, we urgently need new stories—stories that don't mindlessly venerate innovation or reduce writing to mere output. So often, large language models like ChatGPT spit out tired ways of speaking, not simply in their uninspired prose styles, but in their replication of systemic biases of racism, sexism, and ableism. Trained on massive data corpora, much of it stolen from archives of copyrighted works, and coded by inherently flawed humans, these AI tools can give us the best and worst of ourselves. Garbage in, garbage out.

At the same time, AI is an instrumental technology, imagined and deployed for problem-solving ends. From its inception, AI was designed to emulate a specific form of intelligence, one that technology critic Evgeny Morozov likens to a bureaucratic mindset.[8] Forged in the crucible of the Cold War, early AI research pursued machine intelligence with a telos: a singular, goal-oriented drive. One emblematic effort, the General Problem Solver, was developed in 1957 by Herbert A. Simon, Allen Newell, and J.C. Shaw. Its method, "means-end analysis," sought to chart the most efficient path from question to answer—a "universal problem-solver" for structured, knowable problems. But such problems are rare in human life, steeped as it is in emotion, ambiguity, and uncertainty. As computer scientist Terry Winograd observed in 1987, "The techniques of artificial intelligence are to the mind what bureaucracy is to human social interaction."[9] Although modern AI—powered by neural networks and LLMs like ChatGPT—has superseded this rules-based paradigm, it remains framed primarily as a problem-solving tool. Even as today's black-box systems appear more flexible or responsive, they are still driven by a logic of efficiency—a drive toward a telos that rehearses a technocratic vision of progress.

For science and technology scholar Ruha Benjamin, these fundamental problems with AI are obscured by this uncritical march toward progress, one that erases, as Jijingi does in Chiang's story, forms of knowledge that are grounded in history and honed by community. Instead, Benjamin urges us to reimagine AI not as "artificial," but "ancestral intelligence—the insights, experiences, and wisdom that grow under the rubble of progress."[10] It's the "collective know-how, the wisdom that has been cut off from us in part because of who it's associated with."[11]

Writing has never been a solo act. Facilitated by AI, our writing should connect us with our past as much as with our future, with one another as much as ourselves. The best human writing challenges us to open our minds, not close them. We owe it to ourselves to tell stories with this new technology that does the same. If we must write with machines, let it not be to replicate, but to reimagine ourselves.

AFTERWORD

THE ECOLOGY DIGITAL

IMAGINE DANCING IN A MACHINE WHERE EVERY MOVE YOU
make changes the music. Like a conductor, your every potent
gesture unleashes new tunes and fresh melodies. You wave,
the sound soars. You crouch, and it plummets with you. A
kick might activate a drum. As you point, stretch, sway, and
spin, you're dancing to the music, but you're also making it.

This might sound a little bit like a next-generation version
of arcade favorite *Dance Dance Revolution*. But a rudimentary version of this device was built in 1967. The "dancing
suit" was an early project at the Environmental Ecology Lab
(EEL), an experimental collective dedicated to reimagining
the purpose of computers at a formative moment in their

development. The collective first embedded copper wires into elastic bands, which they then sewed into a full-body leotard. As a dancer—in these early tests, an artist named Sansea Sparling—moved, she would stretch the bands, causing the copper to transmit a signal to the "squawk box." Though the prototype apparently didn't sound very good, the suit created a feedback loop between movement and sound, centering the body's idiosyncratic rhythms and improvisations.

Founded in 1967 by ex-psychiatrist Warren Brodey and electrical engineer Avery Johnson, EEL described itself as a "post-industrial laboratory" exploring how technologies might enrich human experience. At the dawn of cybernetics—a field then focused on systems, feedback, and control in machines and living organisms—Brodey and Johnson pursued a countercultural vision. They envisioned computing not as a tool of domination, but communion.

While many in the field were steering computing toward automation and control, often in alignment with Cold War militarism and industrial efficiency, Brodey and Johnson worried this path would lead to mass consumption and corporate capture. Against this tide, they imagined computers that responded to the rhythms of human life. Their work challenged the dominant logic of the industrial and scientific revolutions, which cast nature and the human body as resources to be managed. Instead, they imagined a new paradigm: responsive, personalized computing that expanded human freedom rather than constricting it.

Over the eighteen months of its existence at Lewis Wharf in Boston, Massachusetts, EEL developed tools for expressive, embodied interaction. Computational interactions would allow a person to know themselves better, and to ultimately become more discerning and sophisticated. Their manifesto outlined a vision of computing as an ecological system of humans dynamically communicating with their environments: "We want to put him into control of a responsive dialogue with perceptual grasping of communication about himself inside and outside and in time and space."[1]

The dancing suit was a prototype of this vision. By creating an interactive musical ecology—where movement generated sound, which inspired movement—the suit challenged traditional boundaries between input and output, body and machine. The dancer wasn't merely reacting to the music; she was co-creating it. Under these conditions, improvisation thrived, unconstrained by preset responses. Sutured into a "complex, interactive, responsive system," the dancer's body becomes a dynamic interface within a new ecology of movement, sound, and expression. This vision of the digital body contrasts sharply with today's quantified self—a version of embodiment reduced to data points for optimization and prediction—that structures many of our current encounters with computing.

Evgeny Morozov recounts EEL's story as a radical counterpoint to today's computing, shaped by what he calls "solutionism"—the belief that the right code or device can fix any of humanity's most intractable problems[2]. But solutionism often smooths away complexity, favoring frictionless experiences over deep engagement. Algorithms that recommend music or monitor steps can limit rather than expand our choices, reinforcing the familiar instead of introducing serendipity or fostering growth. And as smartwatches and smart cities do, they reduce the complexity of human bodies and behaviors to data points that can then be crunched by algorithms. For example, fitness trackers might quantify our movement, but they fail to capture the joy or frustration of a particular physical experience, as when my Health app registered my hike around Yosemite National Park as an ascent up a skyscraper.

In Morozov's telling, the kind of responsive, improvisational interactivity envisioned by EEL remains largely unrealized. A decade before Steve Jobs and Steve Wozniak embarked on their more well-known personal computing journey with Apple, EEL imagined machines that could adapt to ever-changing, evolving, and fickle humans in real time. Rather than offer endless variations of the same, their model might guide users to songs—or ideas—they might never have chosen. The dancing suit, in its immediate responsivity, embodies this expansive potential. Its vitality stands in stark contrast to the cold mimicry of Obvious's *Portrait of Edmond de Belamy*, an image about art, but absent of the body that animates it.

Like any of the histories in this book, this paradigm was never inevitable. The trajectory of technological development is steered by the values of its designers, engineers, and architects—and many contemporary digital technologies treat individuality not as a right, but as a product. Our devices cling so closely to our bodies—watching, measuring, tracking—that they could, in theory, enable meaningful personalization. But while today's computers might be technically capable of such nuanced responsivity, they are animated by fundamentally different imperatives. Most prize consumption, and are engineered to optimize it. In practice, the dominant model is extractive: its tools manipulate, capture, and resell our bodies back to us as curated sameness. As they get to know us through surveillance and interaction, algorithmic recommendations serve up not difference, but repetition. Yet, other encounters—mixtape culture, online communities like LambdaMOO, the Chinese programmers who invented new keyboard choreographies—demonstrate alternative ecologies of interaction born from grassroots creativity, not corporate design. These practices reveal a persistent human desire for more participatory and less prescriptive digital spaces, and offer glimpses of richer ways of looking, listening, feeling, and being together.

As the interwoven histories in this book show, humans have always been deeply entangled with their tools. While built to serve us, technologies also shape us. They have replicated and extended our bodies in their senses and faculties, sometimes to our detriment. In many of their current digital forms, they have shrunk our capacities to machine-readable size. But our bodies remain unruly, diverse, and ever-changing. No two bodies move or sense alike. With little legs and keen eyes for hidden treasures, a child navigates a park's pathways for play while her grown-up steers her towards a destination beyond distraction. Though the internet speaks in images, a blind user listens—hearing a website through a screenreader, if it is built to be accessible. People with limited mobility command devices by gaze or breath, rewriting the rules of touch and motion on standard keyboards or touchscreens. And one person's ability may change across time; an able-bodied person may become temporarily or permanently disabled later in life. Age also shapes interaction: devices built primarily for young adults can exclude children or elders with different ergonomic or cognitive needs. Wearables and biometric sensors frequently depend on data calibrated to narrow ranges of body types or skin tones, leaving many users invisible—just as early photography failed darker skin. Culture further contours how someone uses a technology, particularly if it wasn't designed for theirs. Of roughly seven thousand living languages, only fifty to a hundred are supported by major browsers and operating systems and have fonts, typefaces, and keyboards enabling their use across devices. The rest, including Mongolian, Tigrinya, and Kurdish, are digitally disadvantaged, lacking infrastructure that sustains their use in the digital sphere. When technologies assume a one-size-fits-all model, they exclude, constrain, or erase difference. They will either force us to fit or fail.

Rather than fearing new technologies, we need an ecological understanding of our relationships with them—one that sees technology not as separate from nature, but deeply embedded within it. German zoologist Ernst Haeckel coined the term "ecology," originally "oecologia," by joining the Greek *oikos*, meaning "house, dwelling place, habitation" with *logis*, meaning "study of." Ecology names the science of the relationships between living organisms and their environments, between home and world. Millennia of human innovation have transfigured both living beings and the environments they inhabit. We have terraformed rivers, forests, and cities with our machines. As Bernie Krause's soundscape recordings testify, human activity has rewritten nature's symphonies. We can now see, hear, touch, and feel the scale of this impact. A more coherent, empathetic humanity begins by recognizing that we already live inside an engineered ecology. A systems mindset—one that acknowledges the interconnectedness of all things—can guide us. Embracing modern life, amid its gadgets and infrastructure, need not set us in opposition to the natural world. When we design and build through an ecological lens, technology joins the intricate web of life on our planet.

In this light, consider how nonhuman life adapts to human-made worlds. On my college campus, squirrels routinely abscond with entire slices of pizza. Leopards roam the urban development surrounding Sanjay Gandhi National Park in Mumbai, and stray dogs make up to 40 percent of their diet, unexpectedly curbing rabies. In the nineteenth century, as coal pollution filled cities, the English peppered moth's coloration turned darker in response. Many birds use human-made elements as materials to build homes: carrion crows and Eurasian magpies steal and repurpose anti-bird spikes as nest-building material—perhaps the most punk approach of a fellow animal to technology. These creatures show us how technology, even when hostile, can be repurposed for building a home. These adaptations, alongside human-engineered solutions, illustrate the potential for coexisting with and re-shaping technology within altered environments.

Humans, too, have engineered ecological interventions. Wildlife corridors are manmade avenues that connect habitats, often those disrupted by large highways. On Christmas Island, the annual mass migration of millions of red crabs from the forest to the ocean is supported by twenty kilometers of barriers, thirty-one crab underpasses, and a five-mile-high crab crossing at one of the island's busiest roads. In the Netherlands, the Zanderij Crailoo Nature Bridge is the world's largest ecoduct, its eight hundred meters bridging a road, a railway, and a sports park; deer, foxes, rabbits, and moles traverse the bridge. All around us, life is moving. These structures show how technologies can support rather than disrupt life's movement.

Understood ecologically, even digital tools are substrates of human culture. In describing the interdisciplinary communications field of media ecology, media theorist Neil Postman used a biological metaphor to describe the relationship of culture and technology: "In biology, a medium is defined as a substance within which a culture grows; in media ecology, a medium is a technology within which human culture grows, giving form to its politics, ideologies and social organization."[3] Media and technology extend the human senses—toward one another, into and embedded in our environments, and through the networks and ecologies that bind us. Cameras have augmented vision, allowing us to see in darkness, across distance, and through walls, while microphones capture whispers the ear might miss. These tools also broaden the range of the human body, from the roads that stretch our stride across continents to the fiber-optic cables that transmit even our most half-baked thoughts at light-speed across the globe. They can expand our sensory worlds—or just as easily enclose them. They can spawn vastly different possibilities. Satellite photography inspired the overview effect, a sublime understanding of how we are all interconnected on our pale blue dot. But the same technologies also surveil, categorize, and control.

As we are not distinct from the technologies we use, neither are we separate from nature. Technology is also bound up with the past, present, and future of the natural world, the source and site of our endless innovations. We mine the earth for raw materials to build our cars and phones; sound and smoke alike cloud our environments; yet technologies also allow us to measure, and sometimes mitigate, our impact on the natural world. Entwined with the living world, our technologies and digital bodies form new ecologies that make possible and sustain new relationships, new abilities, new sensory capacities, new communities, and new politics. The question is: How do we tend and inhabit these ecologies, to make them home?

One step is to expand how we understand intelligence.

Artist and technologist James Bridle challenges narrow, corporate visions of AI, calling for a broader, ecological view—one that interweaves human, nonhuman animal, plant, and machine ways of knowing. Early AI systems were shaped by bureaucratic logic: classification, control, and the efficient processing of human behavior into data. Bridle contrasts this with the extraordinary intelligences of the natural world: the electrochemical sensitivity of fungi, the problem-solving capacities of slime molds, and the subterranean signals exchanged by plants and trees. Intelligence exists not only in machines or minds, but in bodies, networks, and interactions; it is emergent, embodied, and relational.

This view of intelligence builds on longstanding traditions across the world that recognize intelligence as entangled with land, waters, ancestors, and more-than-human kin. The Zapatistas of Chiapas fight for "a world where many worlds fit," rejecting extraction for coexistence. Yolŋu songlines in Australia's Northern Territory weave land, language, and story into living maps of knowledge. In the Andean cosmovisión, the Quechuan ethic of ayni enshrines reciprocity as a principle of balance between humans, the earth, and the universe. Such ways of knowing remind us that intelligence is not only computed, but cultivated.

Similarly, Ruha Benjamin reimagines AI as ancestral intelligence, rooted in lived experiences across race, gender, age, and ability. This ecological perspective centers the wisdom emerging from diverse embodiments, especially those long ignored and that grow "under the rubble of progress,"[4] and insists that intelligence grows from connection, history, and adaptation—not just calculation.

This shift—from intelligence as control to intelligence as connection—echoes one of this book's central claims: that technologies are not separate from us, but emerge through and alongside our bodies. Most digital technologies assume normative embodiments. If your body doesn't fit, you're told to adjust—or else miss out. But bodies vary in how they look, move, see, hear, sense, and feel. Cultural and physical variations contour every interaction. A more expansive digital embodiment must affirm that diversity. An ecological approach to design begins with the recognition that diversity, interdependence, and imagination aren't problems to solve, but conditions to cultivate. A genuine ecology of technology requires systems thinking, collective design, and communal investment. Communities like LambdaMOO didn't emerge from market incentives; they grew from shared curiosity. To sustain our bodies and our body politic, we need tools that support the integration and interdependence of body, technology, environment, and community. After all, as evolutionary biologist Lynn Margulis once wrote, "Life did not take over the globe by combat, but by networking."[5]

Cultivating a richer digital embodiment means nurturing our own idiosyncratic relationships—with ourselves, our bodies, and the world around us. While data flattens the messiness of human life into digestible bits, we must nourish complexity. We have a choice in how we design and dwell in these ecologies.

In the introduction to this book, I described how technologies become invisible through the union of skilled bodies and elegant tools. Yet, as Mark Weiser observed in the same essay in which he discussed the invisibility of a good tool, most technologies call attention to themselves rather than recede into the background. As an alternative metaphor, he proposed "childhood—playful, a building of foundations, constant learning, a bit mysterious and quickly forgotten by adults. Our computers should be like our childhood—an invisible foundation that is quickly forgotten but always with us, and effortlessly used throughout our lives."[6] Our bodies are made for work as much as they are for play; nurturing the latter makes our tools more inspiring. Reclaiming our bodies for play—bringing curiosity and creativity to our interactions with technology—means honoring this vision. Our technologies ought to spark wonder, support inefficiency, and invite delight—and not simply speed production.

Rather than reflexively embracing or rejecting new

technologies, we must ask: Do they expand or contract our horizons? Do they sustain care, curiosity, and complexity—or reduce us to what can be measured and predicted? How do they shape how we see, move, feel, speak, and connect? The history of our digital bodies shows that the ecologies we create are never neutral. They reflect how we choose to know one another, and how we allow ourselves to be known.

Choreographer William Forsythe once said, "If dance only does what we assume it can do, it will expire."[7] Forsythe saw dance as a form of "motion organization," a means of choreographing not just the body, but the mind—an art of organizing thought, perception, and relation. Known for testing the limits of movement, he understood choreography not merely as the arrangement of steps, but as a structure for sensing and becoming.

Technologies, too, choreograph our lives. Vision machines freeze us. Smart infrastructures herd us. Algorithms script our appetites. They train our gestures, overwrite our rhythms, and codify our repertoire. That's why choreography offers such a powerful lens for understanding the digital body—not as something to be optimized or erased, but as something always becoming, improvising in response to the systems that would contain it.

Few artifacts embody this more vividly than the dancing suit prototyped by EEL: a feedback loop of flesh and circuitry, response and invention. It reminds us that our bodies give every system a pulse, and every interface, a rhythm.

Our bodies do more than perform.

They remember. They resist. They reimagine. They reconfigure.

And they must.

For if our bodies only do what we assume they can do, we will surely expire.

ACKNOWLEDGMENTS

This book emerges from the ecologies that have sustained me. These living networks keep my mind alive, my heart hungry, and my muse well-watered.

While this book attends to the everyday, its ambitions are rooted in the extraordinary. I owe a great debt to the many artists with whom I work on audacious projects that stretch the meaning and capacity of technologies, and whose experiments in embodiment teach me that we don't really know everything a body can do. My deepest appreciation to my colleagues at Leonardo, who hold space for wildly imaginative cultural work that reveals the elasticity of the digital world, and of our shared human potential in it.

This may sound absurd after a book's worth of caution against life danced entirely in code, but algorithmic curators got me through some tough nights of writing. They introduced

222 | VANESSA CHANG

me to new sounds, revived my memories of old ones, and kept me moving at my desk as these chapters found form.

I'm grateful to my editor, Mike Lindgren, for finding me and believing in this book from the beginning, and for his clarity, generosity, and superlative support with shaping it into something people might actually want to read. Thank you to Melville House for trusting his instinct and taking a chance on me, and to their team for shepherding this book into the world with skill and care.

Although this book was composed in a year of late evenings, it has gestated over a decade. Every page of this book bears the fingerprints of the thinkers, readers, and interlocutors who helped shape its contours along the way. At Stanford, I was fortunate to be mentored by Scott Bukatman, Fred Turner, Jisha Menon, and Sianne Ngai; at Vassar, Michael Joyce and Peter Antelyes encouraged me to lean into language.

Special thanks to Lindsey Felt—an incisive thinker, generous collaborator, and dear friend—whose partnership helped awaken me to many of the ideas in these pages. I'm also grateful to the many who lit up when I shared that I was writing this book, and who thoughtfully offered ideas, read excerpts, shared research, or sent artifacts. I'm sure I'll miss some—thank you to Alexis Charles, Ethan Plaut, Antonella Mazzoni, Russell Downham, Mark Bolotin, and Nadav Hochman. To my community at the Castro Writers

Coop: thank you for making it plausible that I could write a book at all.

My family has been the foundation of my life and work. To my brothers—Chris, Jeremy, Johnny, Nick—cheerleaders always: our formative years spent crouched over gaming consoles, battling over the remote, and lounging in front of the TV were the crucible of the body at the heart of this book.

To my parents, Debbie Chang and Chang Wei Chun: thank you for always seeing and embracing me in my pursuits, however leftfield, and for showing me the value of commitment—to people, to work, to play. Thank you, too, for indulging my requests for gadgets; turns out it was all research!

To my partner, Alexander Holt: thank you for your steadfast enthusiasm, your quiet presence, endless cups of tea, and for taking on bedtime duty almost every night so I could keep writing. I owe you so many stories.

And to Avian, whose digital self is still in the making—you are why I dream of technologies that are more just, more joyful, more alive. May the world you inherit be one where the technologies we build nourish more than they consume, and where connection is measured not in bandwidth but in care.

NOTES

INTRODUCTION

1 Mark Weiser. "The World is Not a Desktop." *Perspectives* (Jan 1994), 7–8.

2 Ursula Le Guin. "A Rant about 'Technology.'" https://www.ursulakleguin.com/a-rant-about-technology

3 Ursula Le Guin. "A Rant About 'Technology.'"

CHAPTER 1

1 Richard Sennett. *The Craftsman*. New Haven: Yale University Press, 2009, p. 151.

2 Ada Lovelace. *Scientific Memoirs Selected from the Transactions of Foreign Academies of Science and Learned Society* (1843). Cited in Frank J. Swetz, "Mathematical Treasure: Ada Lovelace's Notes on the Analytic Engine."

3 John Philip Sousa. "The Menace of Mechanical Music." *Appleton's Magazine, Vol. 8* (1906), p. 279.

4 John Philip Sousa. "The Menace of Mechanical Music." *Appleton's Magazine, Vol. 8* (1906), p. 279.

5 John Philip Sousa. "The Menace of Mechanical Music." *Appleton's Magazine, Vol. 8* (1906), p. 284.

6   William Gaddis. *Agapē Agape*. New York: Viking, 2003, p. 30.

7   Thomas Mullaney. *The Chinese Typewriter*. MIT Press, 2018.

8   Berninger et. al, "Early development of language by hand: composing, reading, listening, and speaking connections; three letter-writing modes; and fast mapping in spelling." *Developmental Neuropsychology* 29.1 (2006).

CHAPTER 2

1   Frank Lewis Dyer and Thomas Commerford. *Edison: His Life and Inventions*. New York and London: Harpers & Brothers Publishers, 1910, p. 208.

2   Gary Tomlinson. *A Million Years of Music: The Emergence of Human Modernity*. Zone Books, p. 48.

3   Henry Longfellow. *Hyperion*.

4   Jeffrey Wollock. *The Noblest Animate Motion: Speech, Physiology, and Medicine in Pre-Cartesian Linguistic Thought*, 1997. Amsterdam: John Benjamins.

5   Thomas Edison. "The Phonograph and Its Future" (1878). *The North American Review*, Vol. 126, pp. 533, 527–536.

6   "The Talking Phonograph." *Scientific American* (Dec 22, 1877), p. 384.

7   John Philip Sousa. "The Menace of Mechanical Music." *Appleton's Magazine, Vol. 8* (1906), p. 278.

8   John Philip Sousa. "The Menace of Mechanical Music." *Appleton's Magazine, Vol. 8* (1906), p. 281.

9   B.C. Forbes. "Edison Working on How to Communicate with the Next World." *The American Magazine*, Vol. 90 (Oct 16, 1920), p. 10.

10   Nelson George. *Hip Hop America*. New York: Penguin, 1998, p. 96.

11   Ernst Jentsch. "On the Psychology of the Uncanny" (1906). Tr. Roy Sellars. In *Uncanny Modernity: Cultural Theories, Modern Anxieties*. Eds. Jo Collins and John Jervis. Palgrave Macmillan, 2008.

12   Sigmund Freud. "The Uncanny" (1919) In *The Standard Edition of the Complete Psychological Works of Sigmund Freud* Vol. XVII. London: The Hogarth Press and the Institute of Psycho-analysis, pp. 219–252. P. 219.

13   Alice Wong. "I Still Have a Voice." KQED. May 18, 2023. https://www.kqed.org/perspectives/201601143471/alice-wong-i-still-have-a-voice-2.

14   Kate Crawford. *Atlas of AI: Power, Politics, and the Planetary Costs of Artificial Intelligence*. Yale University Press, 2021, p. 93.

CHAPTER 3

1   Bernie Krause. *The Great Animal Orchestra: Finding the Origins of Music in the World's Wild Places*. Little Brown and Company, 2012.

2   *Animal Music: Sound and Song in the Natural World*. Eds. Tobias Fischer and Lara Cory. Strange Attractor Press, 2015.

3   Camila Zimmerman. "How Regina Music Boxes Mediated Listening Culture." *Local Archives, Global History, UNT Music Library Blog* (Sept 17, 2021). https://blogs.library.unt.edu/music/2021/09/17/how-regina-music-boxes-mediated-listening-culture/.

4   Advertisement for Regina Music Box from *Music Trade Review Magazine*, 1896, The International Arcade Museum.

https://elibrary.arcade-museum.com/Music-Trade-Review.

5   Jonathan Sterne. *The Audible Past*. Duke University Press, 2003, p. 332.

6   Geoffrey O'Brien. *Sonata for Jukebox: Pop Music, Memory, and the Imagined Life*. Counterpoint, 2004, p. 108.

7   Nick Hornby. *High Fidelity*. Riverhead Books, 1996, pp. 62–63.

8   Matias Viegener in *Mix Tape: The Art of Cassette Culture*. Ed. Thurston Moore. Universe Publishing, 2005.

9   Phil Patton. "Humming Off Key for Two Decades." *The New York Times* (July 29, 1999).

10 First iPod commercial, 2001.

11 David Rowell. *The Endless Refrain: Memory, Nostalgia, and the Threat to New Music*. Melville House, 2024.

12 Kyle Chayka. "Why I Finally Quit Spotify." *The New Yorker* (July 31, 2024).

13 Mara Mills. "Do Signals Have Politics? Inscribing Abilities in Cochlear Implants." *The Oxford Handbook of Sound Studies*. Oxford University Press, 2012.

CHAPTER 4

1   Archibald MacLeish. "Riders on Earth Together, Brothers in Eternal Cold." *The New York Times* (Dec 25, 1968).

2   Steve Connor. "Forty years since the first picture of earth from space." *The Independent* (Jan 10, 2009).

3   Joe Moran. "Earthrise: the story behind our planet's most famous photo." *The Guardian* (Dec 22, 2018).

4   Oran W. Nicks, ed. *This Island Earth*. Washington, DC: NASA (1970).

5   Nivedita Bhattacharjee. "Global push for cooperation as space traffic crowds Earth orbit." *Reuters* (Dec 1, 2024).

6   Online Index of Objects Launched into Outer Space. United Nations Office for Outer Space Affairs. https://www.unoosa. org/oosa/osoindex/index.jspx?lf_id.

7   Jenny Odell. "Satellite Collections, 2009–2015." https://www. jennyodell.com/satellite.html.

8   Elizabeth Wallace. *Mark Twain and the Happy Island*. Chicago: A.C. McClurg & Co, 1914.

9   Sebastian Dobson. "Guiding the Sitter." In *Portraiture and Early Studio Photography in China and Japan*. Eds. Luke Gartlan and Roberta Wue. London: Routledge, 2017.

10   Sebastian Dobson. "Guiding the Sitter." In *Portraiture and Early Studio Photography in China and Japan*. Luke Gartlan and Roberta Wue (eds), Routledge, 2017.

11   Trevor Paglen. "Invisible Images (Your Pictures Are Looking at You)." *The New Inquiry* (Dec 8, 2016). https://thenewinquiry. com/invisible-images-your-pictures-are-looking-at-you/.

12   Trevor Paglen. "Invisible Images (Your Pictures Are Looking at You)."

13   Charles Baudelaire. "The Salon of 1859." In *Charles Baudelaire: The Mirror of Art*. Ed. and trans. Jonathan Mayne. London: Phaidon Press, 1955.

CHAPTER 5

1   Aaron Sussman and Ruth Goode. *The Magic of Walking*. Simon and Schuster, 1980.

2   Carol Ann Rinzler. *Leonardo's Foot: How 10 Toes, 52 Bones, and 66 Muscles Shaped the Human World*. Bellevue Literary Press, 2013.

3   Rebecca Solnit. *Wanderlust: A History of Walking*. Penguin Books, 2001, p. 62.

4   Jean-Jacques Rousseau. *The Confessions*. Penguin Classics, 1953, p. 382.

5   Søren Kierkegaard. "Letter to Henrietta Lund," 1847.

6   Friedrich Nietzsche. "Writing with One's Feet (1882)." In *The Gay Science*. Trans. By Walter Kaufmann, New York: Vintage Books, 1974, p. 63.

7   Thomas Clark. *In Praise of Walking*. Moschatel Press, 2004.

8   Robert Macfarlane. *The Old Ways: A Journey on Foot*. Hannah Hamilton UK, 2013, p. 17.

9   Ursula Le Guin. "A Rant about 'Technology.'" https://www.ursulakleguin.com/a-rant-about-technology.

10   Le Corbusier. Vers Une Architecture (Towards an Architecture), 1923.

11   Sadie Plant. *The Most Radical Gesture: The Situationist International In a Postmodern Age*. London and New York: Routledge, 1992, p. 58.

12   Frédéric Gros. *A Philosophy of Walking*. Verso, 2008, p. 296.

13   Rob Kitchin, Tracey P. Lauriault, and Gavin McArdle. "Knowing and governing cities through urban indicators, city benchmarking and real-time dashboards." *Regional Studies, Regional Science, 19 Jan 2015*.

14   Paul McFedries. "The city as system [Technically Speaking]." IEEE Spectrum (April 2014).

15 Taeyoon Choi. "Hello, World!" Avant.org (June 5, 2017). http://avant.org/project/hello-world.

16 "Engineered City." HudsonYardsNewYork.com (2017).

17 Alexa Vaughn. "DeafScape: Applying DeafSpace to Landscape." *Ground Up Journal*, Issue 7 (2018).

18 In *The Architect Says: Quotes, Quips, and Words of Wisdom.* Ed. Laura Dushkes. Princeton Architectural Press, 2012.

CHAPTER 6

1 Julian Dibbell. "A Rape in Cyberspace; or, How an Evil Clown, a Haitian Trickster Spirit, Two Wizards, and a Cast of Dozens Turned a Database into a Society." *The Village Voice*, (1993).

2 Stanley Grauman Weinbaum. "Pygmalion's Spectacles." *Project Gutenberg ebook* (2007 [1949]). https://www.gutenberg.org/files/22893/22893-h/22893-h.htm.

3 Oliver Wendell Holmes. "The Stereoscope and the Stereograph." *The Atlantic* (June 1859), pp. 744, 738–748.

4 Charles Baudelaire. "On Photography" from the Salon of 1859.

5 "Inside Facebook Reality Labs: Wrist-based interaction for the next computing platform." *Tech at Meta* (March 18, 2021). https://tech.facebook.com/reality-labs/2021/3/inside-facebook-reality-labs-wrist-based-interaction-for-the-next-computing-platform/.

6 David J. Chalmers. *Reality+: Virtual worlds and the problems of philosophy.* Penguin Books, 2022.

7 Meta Store, Ray-Ban Meta Wayfarer. https://www.meta.com/ai-glasses/wayfarer/?srsltid=AfmBOoot6EIeeMkXxUta_dEIZpwqlz1jYIopLaTgnE2HInLS0k0RR-KX.

8   Zadie Smith. "Generation Why?" *The New York Review of Books* (Nov 25, 2010).

CHAPTER 7

1   Joseph Weizenbaum. "ELIZA – a computer program for the study of natural language communication between man and machine." *Computational Linguistics/Communications of the ACM* Vol 9. Issue 1 (1966), pp. 36–45.

2   Joseph Weizenbaum. "Contextual Understanding by Computers." *Computational Linguistics* Vol 10. No 8. (Aug 1967): 474-480, p. 474.

3   Joseph Weizenbaum. "ELIZA – a computer program for the study of natural language communication between man and machine." *Computational Linguistics/Communications of the ACM* Vol. 9, Issue 1 (1966): 36-45

4   Joseph Weizenbaum. "Contextual Understanding by Computers." *Computational Linguistics* Vol. 10, No. 8 (Aug 1967): 474-480, p. 478.

5   *Automatic on the Road.* Dir. Lewis Rapkin, Oscillator Media (2018): https://www.youtube.com/watch?v=TqsW0PMd8R0.

6   *Memory and Imagination: New Pathways to the Library of Congress.* Dir. Michael Lawrence and Julian Krainin (1990).

7   "The Persistence of Memory." *Cosmos: A Personal Voyage* (Episode 11).

8   Evgeny Morozov. "The AI We Deserve." *Boston Review* (Dec 4,

     ograd. "Thinking machines: Can there be? Are we?"
     daries of Humanity: Humans, Animals, Machines.

Berkeley: University of California Press, 1991, pp. 198–223.

10 "The New Artificial Intelligentsia." *Los Angeles Review of Books* (18 October 2024). https://lareviewofbooks.org/article/the-new-artificial-intelligentsia/

11 Justin Hendrix. "Imagining 2025 and Beyond with Dr. Ruha Benjamin." *Tech Policy Press* (22 Dec 2024). https://www.techpolicy.press/imagining-2025-and-beyond-with-dr-ruha-benjamin/

AFTERWORD

1 "Manifesto for the Environmental Ecology Lab." https://www.sense-of-rebellion.com/text/manifesto-for-the-environmental-ecology-lab

2 Evgeny Morozov. "The AI we could have had." *Financial Times* (27 June 2024).

3 Kate Milberry. "Media Ecology." *Oxford Bibliographies* (23 May 2012). https://www.oxfordbibliographies.com/display/document/obo-9780199756841/obo-9780199756841-0054.xml

4 Ruha Benjamin. "The New Artificial Intelligentsia." *Los Angeles Review of Books* (October 18, 2024). https://lareviewofbooks.org/article/the-new-artificial-intelligentsia/

5 Lynn Margulis and Dorion Sagan. *Microcosmos: Four Billion Years of Microbial Evolution*. University of California Press, 1997.

6 Marc Weiser, "The World is Not a Desktop." *Perspectives* (Jan 1994), 7–8.

7 Diane Solway. "Is It Dance? Maybe. Political? Sure." *The New York Times* (Feb 18, 2007).

ABOUT THE AUTHOR

DR. VANESSA CHANG is the Director of Programs at Leonardo, the International Society for the Arts, Sciences, and Technology. She has been a lecturer in Visual & Critical Studies at California College of the Arts, lead curator with CODAME Art & Tech, and a SOMArts Curatorial Resident 2019–2020.

INTO
THE BODY
DIGITAL

MEET

VANESSA CHANG